Lecture Notes in Computer Science   9507

Commenced Publication in 1973
Founding and Former Series Editors:
Gerhard Goos, Juris Hartmanis, and Jan van Leeuwen

Editorial Board

More information about this series at http://www.springer.com/series/7409

Pascal Molli · John G. Breslin
Maria-Esther Vidal (Eds.)

Semantic Web Collaborative Spaces

Second International Workshop, SWCS 2013
Montpellier, France, May 27, 2013
Third International Workshop, SWCS 2014
Trentino, Italy, October 19, 2014
Revised Selected and Invited Papers

 Springer

Editors
Pascal Molli
University of Nantes
Nantes
France

Maria-Esther Vidal
Universidad Simón Bolívar
Caracas
Venezuela

John G. Breslin
National University of Ireland Galway
Galway
Ireland

ISSN 0302-9743   ISSN 1611-3349   (electronic)
Lecture Notes in Computer Science
ISBN 978-3-319-32666-5   ISBN 978-3-319-32667-2   (eBook)
DOI 10.1007/978-3-319-32667-2

Library of Congress Control Number: 2016935581

LNCS Sublibrary: SL3 – Information Systems and Applications, incl. Internet/Web, and HCI

Printed on acid-free paper

This Springer imprint is published by Springer Nature
The registered company is Springer International Publishing AG Switzerland

Preface

This volume contains extended papers of work presented at the Second and Third International Workshops on Semantic Web Collaborative Spaces (SWCS), held in 2013 and 2014, respectively.

All of the papers included in this volume went through a two-step peer-review process: They were first reviewed by the Program Committees for acceptance and presentation at the workshops, and then they were extended after the workshop and went through a second review phase. In addition, we invited authors to submit papers on topics related to Semantic Web Collaborative Spaces. All papers were evaluated in terms of technical depth, significance, novelty, relevance and completeness of the references, approach evaluation, and quality of the presentation.

We accepted six out of ten submissions. Our sincere thanks goes to the Program Committee members and external reviewers for their valuable input, and for accepting our invitation to contribute to the review process.

Collaboration between data producers and consumers is a key challenge for facilitating the evolution of the Linking Open Data (LOD) cloud into a participative and updatable LOD cloud. Semantic Web Collaborative Spaces support collaboration among Open Data producers and consumers to publish and maintain Linked Data, as well as to improve their quality. These collaborative spaces include social semantic frameworks such as crowdsourcing tools, semantic wikis, semantic social networks, and semantic microblogs. Collaborative spaces have been developed for different domains, e.g., Health Care, Life Sciences, and E-government.

After our first successful event in Lyon, France, at the 21st International World Wide Web Conference (WWW 2012), our Second International Workshop on Semantic Web Collaborative Spaces was held in Montpellier, France, collocated with the Extended Semantic Web Conference (ESWC 2013), and our Third International Workshop on Semantic Web Collaborative Spaces was collocated with the International Semantic Web Conference (ISWC 2014) in Trentino, Italy.

The Second and Third International Workshops on Semantic Web Collaborative Spaces aimed to bring together researchers from the database, artificial intelligence, and Semantic Web areas, to discuss research issues and experiences in developing and deploying concepts, techniques, and applications that address various issues related to collaborative spaces. Both editions were focused on collaborative data management, models to represent collaborative knowledge and reasoning, tools to interact with SWCS, and associated applications. We are grateful to the ESWC 2013 and ISWC 2014 organizers for their support in making these two meetings successful.

February 2016

Pascal Molli
John G. Breslin
Maria-Esther Vidal

Organization

Workshop Chairs and Organizing Committee

Pascal Molli   University of Nantes, France
John G. Breslin   Insight Centre for Data Analytics, National University of Ireland Galway, Ireland
Maria-Esther Vidal   Universidad Simón Bolvar, Venezuela

Program Committee

Maribel Acosta   AIFB, Karlsruhe Institute of Technology, Germany
Uldis Bojars   University of Latvia, Latvia
Anne Boyer   Lorraine University, France
John Breslin   Insight Centre for Data Analytics, National University of Ireland Galway, Ireland
Tobias Bürger   Capgemini SD&M, Germany
Michel Buffa   University of Nice, France
Amelie Cordier   LIRIS, Lyon University, France
Alicia Diaz   La Plata University, Argentina
Fabien Gandon   Inria Sophia Antipolis, Wimmics, France
Magnus Knuth   Hasso Plattner Institute, University of Potsdam, Germany
Markus Krötzsch   University of Oxford, UK
Christoph Lange   University of Bonn, Fraunhofer IAIS, Germany
Pascal Molli   University of Nantes, France
Claudia Müller-Birn   Freie Universität, Berlin, Germany
Grzegorz J. Nalepa   AGH University of Science and Technology, Poland
Amedeo Napoli   CNRS, LORIA, France
Harald Sack   Hasso Plattner Institute, University of Potsdam, Germany
Hala Skaf-Molli   University of Nantes, France
Sebastian Tramp   University of Leipzig, Germany
Josef Urban   Radboud Universiteit, Nijmegen, The Netherlands
Maria-Esther Vidal   Universidad Simon Bolivar, Venezuela

Sponsoring Institutions

ANR Kolflow Project (ANR-10-CORD-0021), University of Nantes
Science Foundation Ireland (SFI) under Grant Number SFI/12/RC/2289 (Insight)
DID-USB http://www.did.usb.ve

Contents

Challenges in Collaborative Spaces

Challenges for Semantically Driven Collaborative Spaces

Pascal Molli[1], John G. Breslin[2], and Maria-Esther Vidal[3](✉)

[1] University of Nantes, Nantes, France
pascal.molli@univ-nantes.fr
[2] Insight Centre for Data Analytics, National University of Ireland Galway,
Galway, Ireland
john.breslin@nuigalway.ie
[3] Universidad Simón Bolívar, Caracas, Venezuela
mvidal@ldc.usb.ve

Abstract. Linked Data initiatives have fostered the publication of more than one thousand of datasets in the Linking Open Data (LOD) cloud from a large variety of domains, e.g., Life Sciences, Media, and Government. Albeit large in volume, Linked Data is essentially read-only and most collaborative tasks of cleaning, enriching, and reasoning are not dynamically available. Collaboration between data producers and consumers is essential for overcoming these limitations, and for fostering the evolution of the LOD cloud into a more participative and collaborative data space. In this paper, we describe the role that collaborative infrastructures can play in creating and maintaining Linked Data, and the benefits of exploiting knowledge represented in ontologies as well as the main features of Semantic Web technologies to effectively assess the LOD cloud's evolution. First, the advantages of using ontologies for modelling collaborative spaces are discussed, as well as formalisms for assessing semantic collaboration by sharing annotations from terms in domain ontologies. Then, Semantic MediaWiki communities are described, and illustrated with three applications in the domains of formal mathematics, ontology engineering, and pedagogical content management. Next, the problem of exploiting semantics in collaborative spaces is tackled, and three different approaches are described. Finally, we conclude with an outlook to future directions and problems that remain open in the area of semantically-driven collaborative spaces.

1 Introduction

Over the last decade, there has been a rapid increase in the numbers of users signing up to be part of Web-based social networks. Hundreds of millions of new members are now joining the major services each year. A large amount of content is being shared on these networks, and around tens of billions of content items are shared each month. In parallel, similar collaborative spaces are being leveraged in both private intranets and enterprise environments; these collaborative spaces have features mirroring those on the public Web.

© Springer International Publishing Switzerland 2016
P. Molli et al. (Eds.): SWCS 2013/2014, LNCS 9507, pp. 3–9, 2016.
DOI: 10.1007/978-3-319-32667-2_1

With this growth in usage and data being generated, there are many opportunities to discover the knowledge that is often inherent but somewhat hidden in these networks. Web mining techniques are being used to derive this hidden knowledge. In addition, Semantic Web technologies, including Linked Data initiatives to connect previously disconnected datasets, are making it possible to connect data from across various social spaces through common representations and agreed upon terms for people, content items, etc.

In this volume, we will outline some current research being carried out to semantically represent the implicit and explicit structures on the Social Web, along with the techniques being used to elicit relevant knowledge from these structures, and the mechanisms that can be used to intelligently mesh these semantic representations with intelligent knowledge discovery processes.

2 Modelling Collaborative Communities and the Role of Semantics

Semantics represented in ontologies or vocabularies can be used to enhance the description and modelling of collaborative spaces. In particular, annotating data with terms from ontologies is a common activity that has gained attention with the development of Semantic Web technologies. Scientific communities from natural sciences such as Life Sciences have actively used ontologies to describe the semantics of scientific concepts. The Gene and Human Phenotype Ontologies have been extensively applied for describing genes, and there are international initiatives to collaboratively annotate organisms, e.g., the Pseudomonas aeruginosa PAO1 genome[1]. One of the main goals to be achieved to support a precise modelling in collaborative spaces is the development of tools that conduct reasoning on top of existing ontologies and allow for the collaborative annotation of these entities.

In this direction, Goy et al. [3] propose a model to represent views of a data set and different versions of annotations of the data enclosed in the view. Annotations can be personal, allowing for the representation of individual conceptualisations of the portion of the domain represented by the view. Additionally, annotations can be shared in case the annotations can be visible to all members of the collaborative space. The proposed model is part of the project *Semantic Table Plus Plus* (SemT++) [4] which provides a platform to collaboratively describe web resources, i.e., to semantically describe images, documents, videos, or any other resource publicly available on the Web. Existing ontologies are provided to model knowledge about resources represented using the ontology annotations, e.g., the DOLCE[2] and Geographic ontologies[3] provide controlled vocabularies to describe Web resources. The benefits of using personal annotations are illustrated in a use case where documents are collaboratively described following an *authored collaboration policy*. This policy allows an authorised user

---

[1] http://www.pseudomonas.com/goannotationproject2014.jsp.

[2] http://www.loa.istc.cnr.it/old/DOLCE.html.

[3] https://www.w3.org/2005/Incubator/geo/XGR-geo-ont-20071023/.

to delete other user annotations if there is no agreement across all users, while the annotations may remain in the user's local view.

3   Semantic MediaWiki Communities

For over ten years on Wikipedia, templates have been used to provide a consistent look to the structured content placed within article texts (these are called infoboxes on Wikipedia). They can also be used to provide a structure for entering data, so that it is possible to easily extract metadata about some aspect of an entity that is the focus of an article (e.g., from a template field called 'population' in an article about Galway). Semantic wikis bring this to the next level by allowing users to create semantic annotations anywhere within a wiki article's text for the purposes of structured access and finer-grained searches, inline querying, and external information reuse. These are very useful in enterprise scenarios when wikis are used as collaborative spaces where structured data can be easily entered and updated by a distributed community of enterprise users. One of the largest semantic wikis is Semantic MediaWiki, based on the popular MediaWiki system.

In the context of domain-specific applications, Kaliszyk and Urban [5] provide an overview of collaborative systems for collecting and sharing mathematical knowledge. One of the problems to be solved by these systems is the visualisation of formal proofs for people, while providing assistance with the translation of informal statements into formal mathematics. Different problems regarding disagreements between mathematicians, e.g., in terms of the lack of a unique formal language and axiomatic systems, have complicated the development of such formalisation frameworks, necessitating collaborative work like the one implemented in wiki applications such as Wikipedia. These collaborative systems or formal wikis allow for formal verification, the implementation of formal libraries to provide a unified terminology and theorems, versioning of the proofs, semantic resolution of ambiguities, collaborative editing, and other semantic tools. The authors illustrate some of these features in their systems.The Mizar Wiki [1] provides a mathematical library that includes theorems that can be reused in other proofs; additionally, Mizar makes available a proof checker to validate the proposed proofs and to facilitate a peer review process. Furthermore, ProofWeb is a web interface for editing pages and assisting users during theorem demonstration. The authors conclude that there are many challenges to be achieved before collaborative systems in formal mathematics can become a reality. In particular, semantics can play a relevant role in future developments.

A paper by Rutledge et al. [7] introduces a technique to annotate and browse a given ontology based on *Fresnel forms*[4]. Fresnel forms are created via a Protegé plugin that allows users to edit and reuse the ontologies, and also to specify how the semantics should be browsed/displayed (using CSS to create a suitable visualisation for a given data structure). The result is a useful tool that exploits the semantics encoded in an ontology to generate semantic wikis. In particular,

---

[4] http://is.cs.ou.nl/OWF/index.php5/Fresnel_Forms.

the authors apply this technique to bridge Semantic MediaWiki with a browsing interface and semantic forms. Evaluation of this approach was carried out in a case study where the authors discussed the feasibility of implementing their tool via its application to the well-known FOAF ontology, and its usefulness in a Wikipedia infobox-style interface.

Zander et al. [9] tackle the problem of pedagogical content management and evaluate how semantic collaborative infrastructures like semantic wikis can have a positive impact on content reusability and authoring. The novelty of this approach relies on an extension to semantic wikis with knowledge encoded in ontologies like the Pedagogical Ontology (PO) and the Semantic Learning Object Model (SLOM[5]) to enhance expressiveness and interoperability across multiple curricula and pedagogies. Furthermore, the combination of these technologies facilitates the generation of pedagogical knowledge as Linked Data. The effectiveness of the proposed framework is empirically evaluated with a user study and measured using a usability test. Reported results suggest that non-computer specialists are highly satisfied with semantic wikis enhanced with pedagogical knowledge, facilitating the task of creating semantically enriched content for teaching and learning.

4   Exploiting Semantics in Collaborative Spaces

Semantics encoded in ontologies and Linked Data sets can have a positive impact on the behaviour of applications built on top of collaborative spaces. The main challenge to be addressed is how to efficiently extract the knowledge represented in existing Linked Data sets and effectively use this knowledge to enhance existing collaborative spaces. Improvements can be of diverse types, e.g., inferred facts from DBpedia can be used for quality assessment of collaborative space contents; thus, diverse methods to uncover data quality problems should be defined. Additionally, queries against these Linked Data sets have to be crafted in such a way that query answers will provide the insights to discover the missing or faulty content in the collaborative space. Finally, evaluation methodologies are required to determine the quality of the discoveries and to precisely propose changes that will assess high quality content of collaborative spaces.

This special issue compiles two exemplar applications where the benefits of using semantics are clearly demonstrated. First, Torres et al. [8] present BlueFinder, a recommendation system able to enhance Wikipedia content with information retrieved from DBpedia[6]. BlueFinder identifies the classes to which a pair of DBpedia resources belong to, and uses this information to feed an unsupervised learning algorithm that recommends new associations between disconnected Wikipedia articles. The behaviour of BlueFinder is empirically studied in terms of the number of missing Wikipedia connections that BlueFinder can detect; reported results suggest that by exploiting the semantics encoded in DBpedia, BlueFinder is able to identify 270,367 new Wikipedia connections.

---

[5] http://www.intuitel.de/.

[6] http://wiki.dbpedia.org/.

BlueFinder addresses the challenges of uncovering missing content in Wikipedia and designing DBpedia queries to recover missing values. As a future direction, the authors propose extending BlueFinder to allow collaborative validation of these values via crowdsourcing and to include the discovered links in Wikipedia.

Following this line of research, Louati et al. [6] tackle the problem of aggregating heterogeneous social networks, and propose a hybrid graph summarisation approach. The proposed technique extends existing clustering methods such as K-medoids and hierarchical clustering to cluster heterogeneous social networks. The novelty of this approach relies on the usage of attributes and relationships represented in the network, to produce clusters that better fill user requirements. The proposed summarisation techniques implement an unsupervised learning algorithm that uses Rough Set Theory to enhance the precision of known clustering methods: K-medoids and hierarchical. The proposed approach is empirically evaluated in an existing social network data set on US political books[7]. The results suggest that exploiting the semantics encoded in the attributes and relationships positively impacts on the quality of the communities identified by the proposed techniques. The challenges achieved by the authors provide the basis for the development of tools for uncovering patterns of the data that suggest quality problems in the content of the collaborative space. In the future, the authors plan to extend the proposed graph summarisation techniques to dynamically adapt the attributes and relationships used to cluster the input network. This extension will allow for the representation of more general domain restrictions, and in consequence, a more expressive semantically-driven clustering approach.

If Semantic Web technologies have allowed many data providers to publish datasets with their own ontologies, consuming these datasets requires aligning different ontologies and performing entity matching. The Semantic Web community has developed sophisticated tools to tackle this problem, but a generic automatic reliable solution is still an open research problem. Human-machine collaboration can be effective to promote reuse of contextual trusted ontology mappings. This special issue also includes work by Bottali et al. [2] on Okkam, a collaborative tool designed to share ontology mappings. Okkam allows users to disagree on mappings, but also to rank mappings according to their *contextual sharedness*. Consequently, Okkam allows users to quickly find mappings adapted to their usage, knowing their level of agreement for a given particular context.

5   Future Directions

A large research effort has enabled the continuous transformation of content to knowledge, building the Semantic Web from a Web of Documents and the Deep Web. A new major research challenge is to ensure a co-evolution of content and knowledge, making both of them trustable. This means that we must be able to not only extract and manage knowledge from contents, but also to augment contents based on knowledge. Co-evolution of content and knowledge happens in a semantic wiki, but the question is how can we scale this at the levels of the Web of Documents and the Web of Data.

---

[7] http://www-personal.umich.edu/~mejn/netdata/.

Co-evolution of content and knowledge is a powerful mechanism to improve both content and knowledge, but such an evolution has to be powered by a new form of collaboration between humans and AI. From the point of view of people, human-AI collaboration means that we must make formal knowledge and its evolution accessible, usable, editable and understandable, so that people can observe, control, evaluate and reuse this formal knowledge. All research works related to explanations and knowledge revision follow this direction.

In the other direction, from the point of view of computers, human-AI collaboration means that we must be able to take into account the unpredictable behaviours of people who can at any moment in time add or modify content and formal knowledge, with the risk of introducing uncertainty or inconsistency. Research works related to uncertainty management in crowdsourcing and truth-finding algorithms are of interest here.

Bootstrapping the co-evolution of content and knowledge requires that we find new ways for humans and AI to collaborate. Knowledge benefits, i.e. querying, reasoning, fact checking, and truth finding, have to be easily available for any people using the Web. On the other hand, knowledge quality brings with it really challenging questions when deployed, for example, if the mistakes found by people contradict with (and should override) what is formally stated in accepted knowledge bases. Finally, the co-evolution of content and knowledge has to be monitored, ensuring that the overall quality of content and knowledge improves.

Acknowledgements. Pascal Molli has been supported by the ANR Kolflow Project (ANR-10-CORD-0021), University of Nantes. John G. Breslin has been supported by Science Foundation Ireland (SFI) under Grant Number SFI/12/RC/2289 (Insight). Maria-Esther Vidal thanks the DID-USB (Decanato de Investigación y Desarrollo de la Universidad Simón Bolívar).

References

1. Alama, J., Brink, K., Mamane, L., Urban, J.: Large formal wikis: issues and solutions. In: Davenport, J.H., Farmer, W.M., Urban, J., Rabe, F. (eds.) MKM 2011 and Calculemus 2011. LNCS, vol. 6824, pp. 133–148. Springer, Heidelberg (2011)
2. Bortoli, S., Bouquet, P., Bazzanella, B.: Okkam synapsis: connecting vocabularies across systems and users. In: Advances in Semantic Web Collaborative Spaces, Revised Selected Papers of the Semantic Web Collaborative Spaces (SWCS) 2013 and 2014 (2016)
3. Goy, A., Magro, D., Petrone, G., Picardi, C., Segnan, M.: Shared and personal views on collaborative semantic tables. In: Advances in Semantic Web Collaborative Spaces, Revised Selected Papers of the Semantic Web Collaborative Spaces (SWCS) 2013 and 2014 (2016)
4. Goy, A., Magro, D., Petrone, G., Segnan, M.: Collaborative semantic tables. In: Proceedings of the Third International Workshop on Semantic Web Collaborative Spaces Co-located with the 13th International Semantic Web Conference (ISWC), Riva del Garda, Italy, 19 October 2014

5. Kaliszyk, C., Urban, J.: Wikis and collaborative systems for large formal mathematics. In: Advances in Semantic Web Collaborative Spaces, Revised Selected Papers of the Semantic Web Collaborative Spaces (SWCS) 2013 and 2014 (2016)
6. Louati, A., Aufaure, M.-A., Cuvelier, E., Pimentel, B.: Soft and adaptive aggregation of heterogeneous graphs with heterogeneous attributes. In: Advances in Semantic Web Collaborative Spaces, Revised Selected Papers of the Semantic Web Collaborative Spaces (SWCS) 2013 and 2014 (2016)
7. Rutledge, L., Brenninkmeijer, T., Zwanenberg, T., van de Heijning, J., Mekkering, A., Theunissen, J.N., Bos, R.: From ontology to semantic wiki – designing annotation and browse interfaces for given ontologies. In: Advances in Semantic Web Collaborative Spaces, Revised Selected Papers of the Semantic Web Collaborative Spaces (SWCS) 2013 and 2014 (2016)
8. Torres, D., Skaf-Molli, H., Molli, P., Diaz, A.: Discovering wikipedia conventions using DBpedia properties. In: Advances in Semantic Web Collaborative Spaces, Revised Selected Papers of the Semantic Web Collaborative Spaces (SWCS) 2013 and 2014 (2016)
9. Zander, S., Swertz, C., Verdu, E., Perez, M.J.V., Henning, P.: A semantic mediawiki-based approach for the collaborative development of pedagogically meaningful learning content annotations. In: Advances in Semantic Web Collaborative Spaces, Revised Selected Papers of the Semantic Web Collaborative Spaces (SWCS) 2013 and 2014 (2016)

Modeling Collaborative Communities and the Role of Semantics

Shared and Personal Views on Collaborative Semantic Tables

Anna Goy[(⊠)], Diego Magro, Giovanna Petrone, Claudia Picardi, and Marino Segnan

Dipartimento di Informatica, Università di Torino, Torino, Italy
{annamaria.goy, diego.magro, giovanna.petrone, claudia.picardi, marino.segnan}@unito.it

Abstract. The scenario defined by current Web architectures and paradigms poses challenges and opportunities to users, in particular as far as collaborative resource management is concerned. A support to face such challenges is represented by semantic annotations. However, especially in collaborative environments, disagreements can easily rise, leading to incoherent, poor and ultimately useless annotations. The possibility of keeping track of "private annotations" on shared resources represents a significant improvement for collaborative environments. In this paper, we present a model for managing "personal views" over shared resources on the Web, formally defined as structured sets of semantic annotations, enabling users to apply their individual point of view over a common perspective provided in shared workspaces. This model represents an original contribution and a significative extension with respect to our previous work, even being part of a larger project, SemT++, aimed at developing an environment supporting users in collaborative resource management on the Web.

Keywords: Collaborative workspaces · Personal information management · Personal views · Ontology-based content management · Semantic technologies

1 Introduction

Human-computer interaction has greatly changed in the last decade, due to the wide availability of devices and connectivity, and to the consequent evolution of the World Wide Web. In particular, Personal Information Management [1] is facing new challenges: (a) Users have to deal with a huge number of heterogeneous resources, stored in different places, encoded in different formats, handled by different applications and belonging to different types (images, emails, bookmarks, documents,...), despite their possibly related content. (b) Web 2.0 and, more recently, Cloud Computing, in particular the Software-as-a-Service paradigm, have enhanced the possibility of user participation in content creation on the Web, as well as the possibility of resource and knowledge sharing. The interaction of these two aspects provided a great impulse to user collaboration in managing shared resources.

A first step in the direction of providing users with a smart support to face these challenges is represented by semantic technologies, and in particular by semantic

© Springer International Publishing Switzerland 2016
P. Molli et al. (Eds.): SWCS 2013/2014, LNCS 9507, pp. 13–32, 2016.
DOI: 10.1007/978-3-319-32667-2_2

annotations. However, in collaborative environments, where users have to provide coherent semantic annotations of shared resources, disagreements can rise, leading to incoherent, poor and ultimately useless annotations: Either the team works subtractively, keeping only what everyone agrees upon (the resulting annotation being much less useful to everyone), or addictively, keeping everything (which leads to a confusing annotation).

In order to solve these problems, the possibility of keeping track of "private annotations" on shared resources can represent a great improvement for collaborative environments supporting semantic annotation of shared resources. The idea is to provide users with a personal view over shared resources, where their annotations remain stored independently from what other team members do. Personal annotations co-operate with shared ones in the organization and retrieval of shared resources.

In this paper, we present a model for managing such personal views over shared resources on the Web. In particular, in our model, personal views are structured sets of semantic annotations, enabling users to apply their individual point of view over a common perspective provided in shared workspaces. The presented model is part of a larger project, *Semantic Table Plus Plus* (SemT++), which will be briefly described in the following, in order to provide the framework for the personal views model.

The rest of the paper is organized as follows: In Sect. 2 we discuss the main related work, representing the background of our work; in Sect. 3, we briefly present the main features of SemT++, and in Sect. 4 we describe the model supporting personal views. Section 5 concludes the paper by discussing open issues and future developments.

2   Related Work

A survey and a discussion of existing Web-based applications supporting collaboration, including groupware and project management tools or suites, can be found in [2, 3], where T++ was introduced. Strategies for organizing resources have been studied within the field of Personal Information Management, where one of the most relevant research topic is well represented by [4], a book by Kaptelinin and Czerwinski containing a survey of the problems of the so-called *desktop metaphor* and of the approaches trying to replace it. In this perspective, an interesting family of approaches are those grounded into Activity-Based Computing (e.g., [5, 6]), where *user activity* is the main concept around which the interaction is built. A similar approach is proposed in [7, 8], where the authors describe a system supporting collaborative interactions by relying on activity-based workspaces handling collections of heterogeneous resources. Another interesting model discussed in the mentioned book is Haystack [9], a system enabling users to define and manage workspaces referred to specific tasks. The most interesting feature of Haystack workspaces is that they can be personalized.

A research field that is particularly relevant for the approach presented in this paper is represented by studies about systems supporting multi-facets classification of resources. In these systems, resources can be tagged with metadata representing different aspects (*facets*), leading to the creation of bottom-up classifications, collaboratively and incrementally built by users, usually called *folksonomies* [10]. Interesting improvements of tagging systems have been designed by endowing them with semantic capabilities (e.g., [11]), in particular in the perspective of knowledge management [12].

Another important research thread, aiming at enhancing desktop-based user interfaces with semantic technologies is the Semantic Desktop project [13]. In particular, the NEPOMUK project (nepomuk.semanticdesktop.org) defined an open source framework for implementing semantic desktops, aimed at the integration of existing applications and the support to collaboration among knowledge workers, while [14] presents an interesting model connecting the Semantic Desktop to the Web of Data.

A different research area that is relevant for the approach presented in this paper is represented by the studies about resource annotation. The simplest tools supporting annotation enable users to add comments (like sticky notes) to digital documents (e.g., www.mystickies.com, among many others). In these tools, typically, no semantics is associated with user annotations. At the opposite side of the spectrum, we can find NLP-oriented annotation tools, in which annotations are usually labels, referring to a predefined annotation schema, associated with phrases within a document. Some of these systems support collaboration among annotators (e.g., GATE Teamware [15], or Phrase Detectives [16]). Many other approaches provide frameworks for semantic annotation in different domain: Uren et al. [17] include a survey of annotation frameworks with a particular attention to their exploitation for knowledge management, while Corcho [18] surveys ontology-based annotation systems.

As far as the support to shared and personal views on resource annotations is concerned, existing systems tend to focus on a single perspective, sometimes favoring the shared one (e.g., in wikis), sometimes favoring the personal one (e.g., in social bookmarking systems). However, there are some research works which try to integrate shared and personal annotations, like for instance [19], where the need for supporting personal annotations in collaborative environments is motivated. There have been also efforts to provide users with the possibility of adding both private and public notes to digital resources (e.g. [20]). An interesting survey of tools supporting collaborative creation of different types of structured knowledge can be found in [21]: The authors conclude by listing features that users would like to have in collaborative tools supporting knowledge creation, among which "having private and public spaces". With respect to this aspect, the focus of our approach is on structured (ontology-based) *semantic* annotations (mainly describing resource content), and we aim at supporting the integration of both perspectives, enabling users to clearly see at a glance both shared and personal annotations.

3   Overview of SemT++

The SemT++ project is an enhancement of T++, which is described in [2, 3]. The T++ environment allows users to collaboratively manage digital resources. It is based on the metaphor of *tables* populated by *objects* and it has the following main features.

Tables as thematic contexts. In T++, users can define shared workspaces devoted to the management of different activities. Such workspaces are called *tables* and support users in the separated, coherent and structured management of their activities. Users can define new tables, at the preferred granularity level; for instance, a table can be used to manage a work project, to handle children care, or to plan a journey.

Uniform management of heterogeneous objects and workspace-level annotations. Objects lying on tables can be resources of any type (documents, images, videos, to-do items, bookmarks, email conversations, and so on), but T++ provides an abstract view over such resources by handling them in a homogeneous way. Table objects, in fact, are considered as *content items* (identified by a URI) and can be uniformly annotated (by comments and annotations).

Workspace awareness. Workspace awareness is supported by three mechanisms: (a) On each table, a presence panel shows the list of table participants, highlighting who is currently sitting at the table; moreover, when a user is sitting at a table, she is (by default) "invisible" at other tables (*selective presence*). (b) Standard awareness techniques, such as icon highlighting, are used to notify users about table events (e.g., an object has been modified). (c) Notification messages, coming from outside T++ or from other tables, are filtered on the basis of the topic context represented by the active table (see [22] for a more detailed discussion of notification filtering).

User collaboration. An important aspect of T++ tables is that they are collaborative in nature, since they represent a shared view on resources and people: "Tables represent *common places* where users can, synchronously or asynchronously, share information, actively work together on a document, a to-do list, a set of bookmarks, and so on" [2, p. 32]. Table participants, in fact, can (a) invite people to "sit at the table" (i.e., to become a table participant); (b) modify and delete existing objects, or add new ones; (c) define metadata, such as comments and annotations (see below).

T++ has been endowed with semantic knowledge, with the goal of offering users a smarter support to resources management and sharing: SemT++ is, thus, the semantic, enhanced version of T++. In the following we will present SemT++ architecture and prototype (Sects. 3.1 and 3.2) and we will summarize the semantic model implemented in SemT++.

3.1   Architecture

The architecture of SemT++ is shown in Fig. 1.

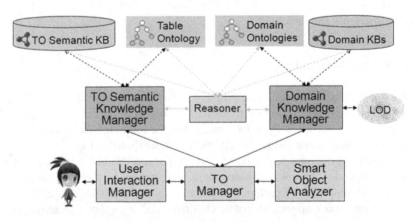

Fig. 1. Sem T++ architecture

The *User Interaction Manager* handles the interaction with users, i.e., the generation of the User Interface and data exchange with the TO Manager.

The *TO (Table Objects) Manager* handles the processes implementing the business logic of the system, namely all the operations taking place on SemT++ tables (e.g., adding/deleting objects, comments, etc.).

The *Smart Object Analyzer* analyzes table objects and extracts information about them; for instance, it finds object parts (e.g., images, links, etc.), it detects the language used and the encoding formats.

The *TO (Table Objects) Semantic Knowledge Manager* manages the semantic descriptions of table objects, stored in the *TO Semantic KB*, and based on the *Table Ontology*, i.e., the system semantic knowledge concerning information objects (see Sect. 3.3).

The *Domain Knowledge Manager* is in charge of the semantic knowledge concerning the content of table objects, stored in the *Domain Knowledge Bases*, and based on one or more *Domain Ontologies*, representing the system semantic knowledge concerning the domain to which table resources refer (see Sect. 3.3). Moreover, the Domain Knowledge Manager handles the connection with Linked Open Data (LOD). The TO Semantic Knowledge Manager and the Domain Knowledge Manager also invokes the Reasoner, when required.

3.2   Prototype and Evaluations

The sketched architecture has been implemented in a proof-of-concept prototype: The backend is a cloud application (a Java Web application deployed on the Google App Engine), while the frontend - implemented by the User Interaction Manager - is a dynamic, responsive Web page, implemented in Bootstrap (getbootstrap.com), using AJAX and JSON to connect to server-side modules (Java Servlets) and to exchange data with the backend.

In the current version of the prototype, the TO Manager relies on Dropbox and Google Drive API to store files corresponding to table objects and Google Mail to handle email conversations. The Smart Object Analyzer exploits a Python Parser Service, which provides the analysis of table objects; currently, it analyzes HTML documents.[1] The Table Ontology and the Domain Ontologies are written in OWL (www.w3.org/TR/owl-features); the TO Semantic Knowledge Manager and the Domain Knowledge Manager exploit the OWL API library (owlapi.sourceforge.net) to interact with them, while the Reasoner is based on Fact++ (owl.cs.manchester.ac.uk/tools/fact). The knowledge bases, containing assertions concerning the semantic description of table objects, are stored in a Sesame (rdf4j.org) RDF triplestore, accessed by the TO Semantic Knowledge Manager and the Domain Knowledge Manager through Sesame API.

---

[1] Besides being a very common format, quite easy to parse, HTML poses interesting challenges to the semantic modeling, since it introduces a further layer – the HTML encoding – between the "digital object", encoded for instance in UTF-8, and the information content representing the Web page itself. We are extending the Smart Object Analyzer functionality in order to analyze other formats.

In order to evaluate our approach, we implemented a testbed case of domain knowledge, i.e., we instantiated a Domain Ontology on geographic knowledge, while the Domain Knowledge Manager connects to GeoNames Search Web Service (www.geonames.org/export), as a significant example of LOD dataset. Further details about this choice can be found in [23].

We evaluated the most important functionalities of SemT++ through some user tests.

Goy et al. [3] report the results of a user evaluation of T++ in which we asked users to perform a sequence of pre-defined collaborative tasks (communication, resource sharing, and shared resources retrieval) using standard collaboration tools (like Google Drive and Skype) and using T++. The results demonstrate that performing the required tasks with T++ is faster and it increases user satisfaction.

Goy et al. [24] present the results of an empirical study about the impact of the semantic descriptions. Potential users of SemT++ were asked to go through a guided interaction with SemT++, aimed at selecting table objects on the basis of multiple criteria offered by their semantic descriptions; then, participants answered a post-test questionnaire. The analysis of users' answers confirmed our hypothesis: An environment supporting the uniform management of different types of resources (documents, images, Web sites, etc.) and the possibility of selecting them by combining multiple criteria (among which resource content) is highly appreciated, and contributes to provide an effective access to shared resources and a less fragmented user experience.

Goy et al. [25] describe a qualitative user study aimed at analyzing user requirements and defining the model supporting collaborative semantic annotation of table objects. Participants were organized into small groups and were asked to collaboratively annotate shared resources, by providing annotations in the form of tags, describing the resources content. Each group experimented different collaboration policies (unsupervised vs supervised, with users playing different roles). At the end of the experiment, participants were asked to fill in a questionnaire where they had to rate their experience and express their opinions about different features. From this user study, we extracted a set of guidelines for designing the model handling collaborative semantic annotation of table objects in SemT++.

3.3   Semantic Model

The core of the approach used in SemT++ to provide users with a flexible and effective management of shared heterogeneous resources is its semantic model, represented by the ontologies and knowledge bases introduced above.

In the following, we will briefly describe it, before concentrating on the focus of this paper, i.e. the framework supporting *personal views* over annotations of shared resources. A more detailed description of SemT++ semantic model can be found in [23, 24].

The Table Ontology models knowledge about information resources. It is grounded in the Knowledge Module of O-CREAM-v2 [26], a core reference ontology for the Customer Relationship Management domain developed within the framework provided by the foundational ontology DOLCE (Descriptive Ontology for Linguistic and Cognitive Engineering) [27] and some other ontologies extending it, among which the Ontology of Information Objects (OIO) [28]. The Table Ontology enables us to describe resources lying on tables as *information objects*, with properties and relations. For instance: A table object (e.g., a document) can have parts (e.g., images), which are in turn information objects; it can be written in English; it can be stored in a PDF file, or it can be a HTML page; it has a content, which usually has a main topic and refers to a set of entities (i.e., it has several objects of discourse). Given such a representation, reasoning techniques can be applied, in order to infer interesting and useful knowledge; for example, if a document contains an image of the Garda lake, probably the document itself is (also) about the Garda lake.

The most relevant class in the Table Ontology is *InformationElement*, with its subclasses: *Document, Image, Video, Audio, EmailThread*, etc. All table objects are instances of one of them. In order to characterize such classes, we relied on a set of properties (some of them inherited from O-CREAM-v2) and a language taxonomy defined in O-CREAM-v2, representing natural, formal, computer, visual languages. A complete account of such properties is out of the scope of this paper; in the following we just mention the most important ones:

- *DOLCE: part(x, y, t)* - it represents relations such as the one between a document (*x*) and an image or a hyperlink (*y*) included in it, holding at time t.[2]
- *specifiedIn(x, y, t)* - it represents relations such as the one between a document (*x*) and the language (*y*) it is written in (e.g., Italian), holding at time *t*.
- *hasAuthor(x, y, t)* - it represents the relation between an information element (*x*) and its author (*y*), holding at time *t*.
- *hasTopic(x, y, t)* - it represents the relation between an information element (*x*) and its main topic (*y*), holding at time *t*.
- *hasObjectOfDiscourse(x, y, t)* - it represents the relation between an information element (*x*) and the entity (*y*) it "talks about", holding at time *t;* it is a subproperty of *OIO: about.*
- *identifies(x, y, t)* - it represents, for instance, the relation between a hyperlink (*x*) and the resource (*y*) it points to, holding at time *t*.

As we mentioned above, we chose commonsense geographic knowledge as a testbed example of system domain competence. This knowledge is represented by a Geographic Ontology coupled with a Geographic KB, containing information retrieved from GeoNames, a huge, open geographical database containing over 10 million geographical entities. For each topic and object of discourse used to describe the content of resources on a given table, the Domain Knowledge Manager searches for corresponding GeoNames entities. If the search result is not void, after a possible disambiguation phase (currently done by the user), the instance representing that topic/object of discourse is classified in the Geographic Ontology (see [23] for more details about the geography testbed).

Figure 2 graphically depicts a simplified example of the semantic description of a table object, i.e., a Web page, with the following properties: The author is a company for touristic promotion of Garda lake (*VisitGarda inc.*); it is written in English; it is encoded in HTML (specifically UTF-8/HTML5); it contains a figure (*image1*) and a

---

[2] Parameter *t*, representing time, is omitted in the OWL version of the Table Ontology.

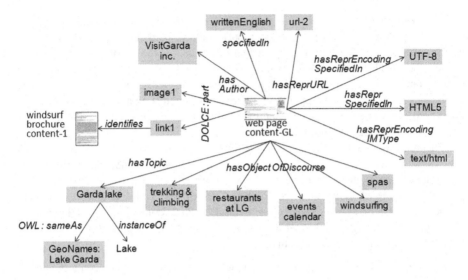

Fig. 2. Simplified example of the semantic description of a table object (Web page)

link (*link1*) to a brochure (*windsurf-brochure-content-1*); its main topic is the Garda lake, and it talks about trekking and climbing, restaurants and events in the lake area, windsurfing, and spas. Since the topic (*Garda lake*) is a geographic entity, the Geographic Knowledge Manager found the corresponding GeoNames entity (*GeoNames: Lake Garda*), and classified it as an instance of the class *Lake* (in the Geographic Ontology).

An important fragment of the proposed semantic model refers to *candidate relationships*: The Reasoner, on the basis of axioms like the following one, can infer *candidate* features, mainly from included objects:

$$InformationElement(x) \land DOLCE : part(x, z, t) \land hasObjectOfDiscourse(z, y, t)$$
$$\rightarrow hasCandidateObjectOfDiscourse(x, y, t)$$

For example, the Reasoner can infer that the city of Trento (y) is a candidate object of discourse of a document (x) – at time t – from the fact that the document itself (x) includes a video (z) about Trento (y) - at time t.

When the Reasoner infers such candidate relationships, the system asks the user for a confirmation: If (and only if) the user confirms, for instance, that *Trento* is actually an object of discourse of the *document*, then a new relation *hasObjectOfDiscourse(doumentc, Trento, t)* is added to the knowledge base.

The semantic descriptions of table objects based on the model just sketched enable table participants to specify and combine multiple criteria in order to select objects on a table. For example, to get all email threads talking about Garda lake windsurfing (i.e., having it as main topic), a user can specify the following parameters: *topics = {Garda_lake_windsurfing}*, *types = {emailThread}*. The current User Interface enabling such a functionality is a sequence of simple Web forms, and is described in [24]. Moreover, the user could provide more general queries, such as asking for all resources talking about lakes, thanks to the facts that topics and objects of discourse (e.g., *Garda_lake*) are represented as instances of classes in the Geographic Ontology (e.g., *Lake*). Finally, specific information about topics and objects of discourse (characterizing the content) retrieved from LOD sets, such as GeoNames, can provide table participants with a sort of "explanation" about such property values.

3.4   Collaborative Semantic Annotation

When a new object is added to a table, or when an existing one is modified (e.g., when a table participant includes a new image in it), the corresponding semantic representation is created or updated. Consider the new object case (the update case is analogous): The semantic representation is created (updated) as follows:

- The Smart Object Analyzer sets some property values (e.g., *DOLCE: part*, properties related to encoding formats), and proposes candidate values for other properties (e.g., *specifiedIn*, *hasAuthor*).
- The Reasoner, invoked by the Semantic Knowledge Manager, proposes other candidates (e.g., topics and objects of discourse).
- The user can confirm or discard candidate values, and add new ones.

The property values that mostly depend on the personal view of each table participant about the content (and scope) of table resources are *hasTopic* and *hasObjectOfDiscourse*. In the case of *hasTopic*, table participants need to agree on a single value expressing the main topic of the resource; in the case of *hasObjectOfDiscourse*, they need to agree on a set of values.

In order to facilitate collaboration and the achievement of an agreement, we designed and implemented a collaboration model for handling semantic annotations on table resources, based on the outcomes of a qualitative user study. Both the study and the implemented model are described in [25]. In the following, we briefly summarize the most relevant features of SemT++ collaboration model.

SemT++ provides three alternative collaboration policies: (1) *Consensual*, where the editing of semantic descriptions of table objects is always possible for all participants in a totally "democratic" way. (2) *Authored*, where the final decision about the semantic annotation of a table resource is taken by its creator (owner). (3) *Supervised*, where the final decision about the semantic annotation of a table resource is taken by the table supervisor. SemT++ enables the table creator to select the collaboration policy to be applied for handling the collaborative process of building semantic descriptions of table objects. Moreover, a resource semantic description can also be simply marked as "approved" by participants (see the checkbox at the bottom of Figs. 4 and 5). Finally, SemT++ explicitly encourages table participants to use the communication tools available on the table, i.e., the Blackboard for posting asynchronous messages, the Chat for synchronous communication, and free-text Comments which can be attached to table objects, prompting them to add an optional comment whenever they edit the annotation (see also [29]).

4   Personal Views over Shared Resources

One of the most interesting results of the user study discussed in [25] is that many users said they would be interested in the possibility of having *personal annotations*, visible only to the author of the annotation itself. Users explained that they would see this functionality as particularly useful for the search and retrieval of table resources based on tags describing their content. Moreover, the importance of supporting personal annotations in collaborative environments has also been claimed in the literature; see, for instance, [19].

Starting from this suggestion, we designed a new functionality for SemT++ enabling table participants to keep their own perspective over table resources. Results observed during the empirical study suggest that disagreements about semantic annotations of shared resources are quite common, especially as far as resource content is concerned. This fact reflects the intuitive common experience that people often have different interpretations of the "meaning" of an information object (e.g., a document, a movie), and it can be quite difficult to reach a consensus about the list of issues/topics it is about. The availability of *personal views* over semantic annotations, within a collaborative environment, represents an advantage, since users can maintain "private" annotations (possibly sources of disagreement) over shared resources.

SemT++ view management resorts on semantic annotations (i.e., semantic properties) of an information object to collaboratively handle resources on the Web. To illustrate a semantic annotation, suppose that x represents an information object, y corresponds to an entity, and t represents a timestamp, then the property assertion *hasObjectOfDiscourse(x, y, t)* is a semantic annotation of the object x (see Sect. 3.3). From the point of view of a single table participant (tp), each semantic annotation can be as follows:

- Case A: visible to all table participants (including tp), but not explicitly "liked" by tp;
- Case B: visible to all table participants (including tp) and "liked" by tp;
- Case C: visible only to tp.

Given a SemT++ table, a *shared view* corresponds to the set of all semantic annotations that fall in cases A or B, while a *personal view* is the set of all semantic annotations in cases B or C. Moreover, we say that a table participant *likes* an annotation to mean that she agrees with it and thus she has explicitly imported it from the *shared view* into her *personal view*. When a table participant (tp) adds an annotation to a table object, she can decide to add it to the *shared view* (in this case it is also automatically marked as *liked* by tp, ending up in case B), or only to her *personal view* (case C). An annotation initially added to the *personal view* (C) can later on be shared (i.e., moved to case B). Moreover, as already mentioned, tp can mark as *liked* annotations added by other participants and belonging to the *shared view* (which means moving an annotation from case A to case B). Finally, tp can delete annotations:

- If the annotation belongs to case A, it is deleted from the *shared view* (see [25] for a detailed account of collaboration policies handling decisions about annotation removal), but maintained in the *personal views* of users who *like* it;

- If the annotation belongs to case B, it is deleted from the *shared view*, maintained in the *personal views* of users who *like* it, but deleted from *tp personal view*;
- If the annotation belongs to case C, it is simply deleted from *tp personal view*.

To support workspace awareness, *tp* can see the author of an annotation and users who like it, by right-clicking on it.

As we mentioned above, the availability of shared and personal views over semantic annotations is particularly interesting for annotations representing the content of table resources, i.e., the *hasTopic* and *hasObjectOfDiscourse* properties. However, the mechanism is available for all "editable" annotations, i.e. property values that are ultimately set by users (e.g., *specifiedIn*, representing the natural languages used in a document, or *hasAuthor*, representing document authors). Property values that are set by the system, e.g., mereological composition and encoding formats (see Sect. 3.4), are instead automatically assigned to case B for every table participant: They automatically belong to the shared view and everybody likes them, i.e., they belong also to all personal views.

In the following we will provide a usage scenario presenting shared and personal views on SemT++ tables (Sect. 4.1), focusing on the most interesting property with respect to this issue, i.e., *hasObjectOfDiscourse*; then, we will describe the underlying mechanisms enabling views management (Sect. 4.2), and sketch our evaluation plan (Sect. 4.3).

4.1   Usage Scenario

Aldo is a volunteer working for Our Planet, a NGO for environment safeguard. Some months ago he created a table (named *Our Planet*) to collaborate with a small team of other local volunteers. Now Aldo has to write an article for an online local newspaper, discussing the situation of the Champorcher mule track: To this purpose, he needs to retrieve information about that topic, available on the *Our Planet* table. He thus asks for the *topics* present on the table, selects *Champorcher mule track*, and gets the list of table objects having it as main topic. Among the results there is a resolution by the Municipality of Champorcher concerning an enlargement project, and two images of Champorcher surroundings. After reading the Municipality's resolution, Aldo creates a new table object (an HTML document, since the article will be published online), writes some text in it, adds one of the just retrieved images, and includes a link to the resolution.

When Aldo decides to leave the table and clicks on the "save&update" button, the table asks him to take a look at the annotations describing his article. Since the resource is new, only the following properties have already been set by the system (see Sect. 3.4):

- *Type*: *Document* (i.e., the class in the Table Ontology to which Aldo had assigned the resource when created);
- *Format*: *HTML* (referring to formal properties such as *hasRepresentationSpecifiedIn(x, y, t)*);

- *Contains*: link to *resolution, image23* (referring to the formal property *DOLCE: part*(*x, y, t*)).

For the following properties, SemT++ suggests some candidate values:

- *Main topic* (referring to the formal property *hasTopic*(*x, y, t*));
- *Objects of discourse* (referring to the formal property *hasObjectOfDiscourse*(*x, y, t*)) - candidate objects of discourse are visible in Fig. 3;
- *Language* (referring to the formal property *specifiedIn*(*x, y, t*));
- *Authors* (referring to the formal property *hasAuthor*(*x, y, t*)).

The window displaying the properties of the new table object (Aldo's article) is shown in Fig. 3: The panel referring to objects of discourse is open; immediately below there is a text field where the user can add new values (objects of discourse in this case). Moreover, the panel with SemT++ suggestions (i.e., candidate objects of discourse) is available.

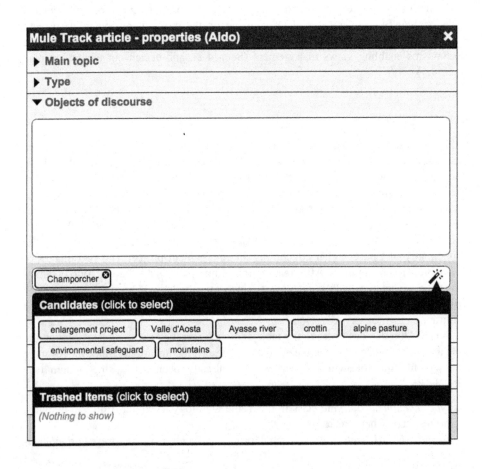

Fig. 3. SemT++ UI displaying semantic annotation of a new table resource

Aldo can select suggested values by clicking on them, or write new ones in the text field, with the support of an autocompletion functionality, based on the values already available on the table.

In our scenario, Aldo - besides confirming the suggested main topic, language and author - selects the following objects of discourse suggested by the system: *Champorcher, enlargement project, Ayasse river*; it also adds *Mont Avic park* and *demonstration 14/5*. His choices can be seen in Fig. 4, which represents the perspective of another table participant, Maria, on the same resource. When adding values to a property, whether new values or suggested ones, the user can choose to add them only to her/his personal view or also to the shared view (via the drop-down menu next to the + button, Fig. 4). In our scenario, Aldo has added all values to the shared view, thus making them visible to all table participants (including Maria).

When, later on, Maria sits at the *Our Planet* table, she reads Aldo's article and then accesses its properties (Fig. 4). By right-clicking on a value, Maria can see the author of the annotations (Aldo in our current example) and users who agree with (*like*) them (again, only Aldo in this case). She decides to mark as *liked* some values (i.e., *Champorcher,*

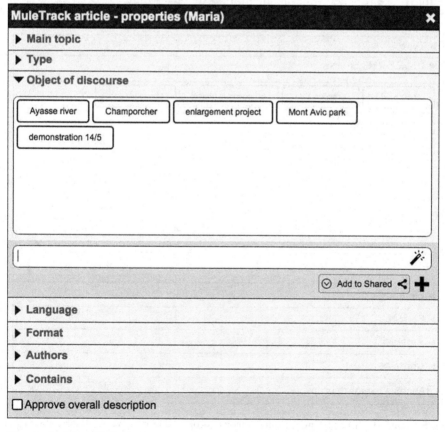

Fig. 4. SemT++ UI displaying semantic annotation of an already annotated table resource

enlargement project, Mont Avic park) and to add to the shared view a couple of new objects of discourse, i.e. *Our Planet* and *Legambiente VdA* (since, according to her opinion, in the article Aldo also talks about the activity of these two NGO). Moreover, she leaves a comment explaining her point of view. By adding opinions to Maria's comment, other table participants can discuss the opportunity of having such objects of discourse in the shared view.

The *Our Planet* table implements an *authored collaboration policy* (see Sect. 3.4), enabling Aldo (who is the author of the resource in focus) to make the final decision about property values. Thus, on the basis of the opinions expressed by table participants, Aldo decides to stop the discussion and to delete the two annotations added by Maria. He writes a final comment to explain his decision.

The two annotations, deleted from the shared view, remain available in Maria's personal view, so that she will be able to use them to select table objects in the future. Maria's final view on Aldo's article is shown in Fig. 5: Bold face boxes indicate annotations belonging to the shared view; a small heart means that Maria likes the

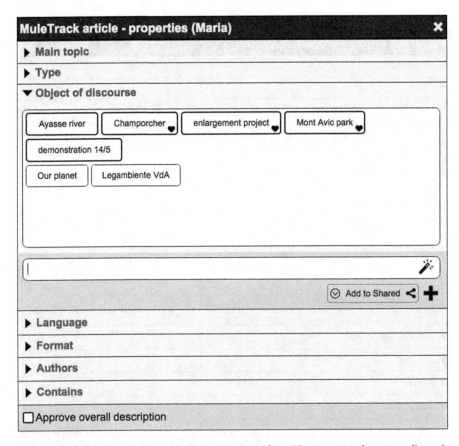

Fig. 5. SemT++ UI displaying semantic annotation of a table resource after some discussion among users

annotations (i.e., hearts mark shared annotation also belonging to Maria's personal view); thin face boxes represent "private" annotations, belonging only to Maria's personal view.

4.2   Semantic Model Enabling Personal Views

As we mentioned in Sect. 3.1, the TO Semantic Knowledge Manager is in charge of managing the semantic descriptions of table objects (stored in the TO Semantic KB and based on the Table Ontology). In order to handle personal and shared views, following the model described above, we implemented a submodule, called *View Manager*, having the role of handling views. Figure 6 shows the new internal architecture of the TO Semantic Knowledge Manager.

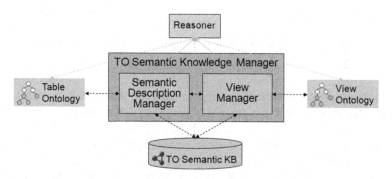

Fig. 6. Internal architecture of the TO Semantic Knowledge Manager of SemT++

The View Manager is endowed with specific knowledge represented by a *View Ontology*, according to which:

- A *context* (an instance of the *Context* class) represents a context in which a given set of assertions holds.
- A *view* (an instance of the *View* class) is a particular type of context (*View* is a subclass of *Context*).
- Assertion sets (instances of the *AssertionSet* class) are linked (by the *holds_in* relation) to contexts (and views).

An assertion set is represented by a named-graph, grouping assertions in the *TO Semantic KB triplestore* (the IRI of the *AssertionSet* instance is the name of the corresponding named-graph).

   When a new table is created, the View Manager creates the following entities, stored in the TO Semantic KB, and depicted in Fig. 7:

- A new *Context* instance (e.g., *infoObjectContext*), representing the "table context", where assertions that are valid in all views (e.g., assertions made by the system) hold;
- A new *AssertionSet* instance (e.g., *infoObjectAssertionSet*) linked to *infoObject Context* by the *holds_in* relation;

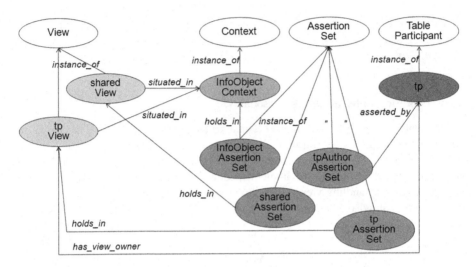

Fig. 7. The (simplified) semantic model for handling shared and personal views on a table in SemT++

- A new *View* instance (*sharedView*), representing the shared view on that table;
- A new *AssertionSet* instance (*sharedAssertionSet*) linked to *sharedView* by the *holds_in* relation;
- A relation (*situated_in*) between *sharedView* and *infoObjectContext*, which enables the shared view to inherit all the assertions holding in the table context; in this way, for example, assertions made by the system automatically hold in the shared view.

Moreover, for each new table participant, the View Manager creates:

- A new instance (*tp*) of the *TableParticipant* class;
- A new *View* instance (*tpView*), representing *tp personal view* on the table, linked to *tp* by the *has_view_owner* relation;
- A relation (*situated_in*) between *tpView* and *infoObjectContext*, which enables the personal view to inherit all the assertions in the table context, so that, for example, assertions made by the system automatically hold in all personal views.
- A new *AssertionSet* instance (*tpAssertionSet*) linked to *tpView* by the *holds_in* relation;
- A new *AssertionSet* instance (*tpAuthorAssertionSet*) linked to *tp* by the *asserted_by* relation.

The system, on the basis of both the View Ontology and the Table Ontology, guarantees that each view (i.e., the shared one and all personal views) is *consistent*, thus enabling the Reasoner to run on each view, in order to make inferences, such as - for instance - those supporting the suggestion of candidate values (see Sect. 3.3).[3]

---

[3] From the implementation point of view, this requires that all the assertions in the triplestore are loaded in the OWL knowledge base (together with the ontology).

The described model enables SemT++ to support the usage scenario sketched in Sect. 4.1. Since the *Our Planet* table had been created months before, the model depicted in Fig. 7 is already up. The following schema summarizes what happens for each step in the scenario: The left column represents the user actions (triggering events), the right column the consequent system action (in this column, the arrow represents a cause-effect relation):

Aldo clicks on the "save&update" button	The properties automatically set by the system are added to the *infoObjectAssertionSet* ↓ The TO Semantic Knowledge Manager runs the Reasoner and calculates candidates (thanks to the *DOLCE: part* relation holding between the image included in the article and the article itself, as well as between the link to the resolution included in the article and the article itself)
Aldo selects some candidate values and adds some new ones, deciding to add all of them to the shared view	The corresponding assertions are added to the *sharedAssertionSet*, to *AldoAssertionSet*, and to *AldoAuthorAssertionSet*
Maria looks at the author of an annotation	The TO Manager gets from the TO Semantic Knowledge Manager the reference to Aldo, retrieved by the View Manager from the *asserted_by property* linking *AldoAuthorAssertionSet* to its author (Aldo)
Maria looks at users who agree with an annotation (i.e., table participants who *like* it)	The TO Manager gets from the TO Semantic Knowledge Manager the reference to Aldo, retrieved by the View Manager from: (i) The *has_view_owner* property linking *AldoView* (i.e., Aldo's personal view) to Aldo; (ii) The *holds_in* property linking Aldo's personal view to *AldoAssertionSet*
Maria marks as *liked* some values	The corresponding assertions are added to *MariaAssertionSet*
Maria adds a couple of new objects of discourse to the shared view	The corresponding assertions are added to the *sharedAssertionSet*, to *MariaAssertionSet*, and to *MariaAuthorAssertionSet*
Aldo decides to delete Maria's annotations from the shared view	The corresponding assertions are deleted from the *sharedAssertionSet* (they remain available in *MariaAssertionSet* and in *MariaAuthorAssertionSet*)

In order to display the properties of Aldo's article (depicted in Fig. 5), the TO Manager gets from the TO Semantic Knowledge Manager all the information needed to display bold face boxes (for shared annotations), small hearts (for shared annotations also belonging to Maria's personal view), and thin face boxes (for personal annotations).

4.3   Future Evaluation Plan

Although the proposed approach is grounded in the results of a qualitative user study (described in detail in [25]), we plan to conduct a comparative evaluation, aimed at verifying the benefits of the availability of a personal point of view on resource annotations, in the context of collaborative workspaces. The evaluation will be carried out with the same methodology as the preliminary user study, in order to be able to compare the results. Therefore, three groups of users will be asked to collaborative annotate a few resources in a pre-defined scenario. Each group will repeat the experience twice, experimenting with different policies (*supervised*, *authored* and *consensual*). While in the initial study users performed the task with Google Documents, in the conclusive evaluation they will have the chance to use the SemT++ environment with the shared/personal views enhancement.

The goal of the evaluation is to determine to which degree limitations and difficulties perceived in the preliminary test are successfully overcome by our approach to collaborative annotation, and in particular by the availability of personal and shared views. This goal will be achieved by providing participants with a questionnaire they will have to fill in after performing the assigned tasks. The research questions we will address by means of the evaluation are listed below; the quality of the experience is qualitatively measured by the following parameters: Interest, engagement, usefulness, and difficulty. With respect to the initial scenario, where personal views were not available, the evaluation aims at answering the following *research questions*:

- Is the *experience* of collaborative annotation improved?
- Are the participants better satisfied with the *final annotations*?
- Are the participants better satisfied with the *collaboration and communication* within the group?

5   Conclusions and Future Work

In this paper we presented a model for handling both shared and personal views on Web resources. The presented approach is part of the SemT++ project, aimed at providing users with a collaborative environment for the management of digital resources in collaborative thematic workspaces. In the paper we described how personal views are handled in SemT++ as structured sets of semantic annotations, represented by semantic assertions grouped into named-graphs within a triplestore knowledge base, and supported by an ontology-based representation, modeling contexts, views, etc.

The approach described in this paper can be exploited to support other improvements of the user experience. For example, as far as users agree on making their personal views visible to other people, such views could represent an interesting source of knowledge about users interpretation of the annotated resources. Moreover, since the reasoner provides an explanation for the inconsistencies it detects, we will study the possibility of exploiting this information for interactively supporting users in solving such inconsistencies. Finally, we are studying the impact of making tables public, together with their shared and personal views: This is an interesting perspective that deserves a deeper analysis.

References

1. Barreau, D.K., Nardi, B.: Finding and reminding: file organization from the desktop. ACM SIGCHI Bull. **27**(3), 39–43 (1995)
2. Goy, A., Petrone, G., Segnan, M.: Oh, no! Not Another Web-based Desktop! In: 2nd International Conference on Advanced Collaborative Networks, Systems and Applications, pp. 29–34. XPS Press, Wilmington, DE (2012)
3. Goy, A., Petrone, G., Segnan, M.: A cloud-based environment for collaborative resources management. Int. J. Cloud Appl. Comput. **4**(4), 7–31 (2014)
4. Kaptelinin, V., Czerwinski, M. (eds.): Beyond the Desktop Metaphor. MIT Press, Cambridge (2007)
5. Bardram, J.E.: From desktop task management to ubiquitous activity-based computing. In: Kaptelinin, V., Czerwinski, M. (eds.) Beyond the Desktop Metaphor, pp. 223–260. MIT Press, Cambridge (2007)
6. Voida, S., Mynatt, E.D., Edwards, W.K.: Re-framing the desktop interface around the activities of knowledge work. In: UIST 2008, pp. 211–220. ACM Press, New York, NY (2008)
7. Geyer, W., Vogel, J., Cheng, L., Muller M.J.: Supporting activity-centric collaboration through peer-to-peer shared objects. In: Group 2003, pp. 115–124. ACM Press, New York, NY (2003)
8. Muller, M.J., Geyer, W., Brownholtz, B., Wilcox, E., Millen, D.R.: One-hundred days in an activity-centric collaboration environment based on shared objects. In: CHI 2004, pp. 375–382. ACM Press, New York, NY (2004)
9. Karger, D.R.: Haystack: per-user information environments based on semistructured data. In: Kaptelinin, V., Czerwinski, M. (eds.) Beyond the Desktop Metaphor, pp. 49–100. MIT Press, Cambridge (2007)
10. Breslin, J.G., Passant, A., Decker, S.: The Social Semantic Web. Springer, Heidelberg (2009)
11. Abel, F., Henze, N., Krause, D., Kriesell, M.: Semantic enhancement of social tagging systems. In: Devedžić, V., Gašević, D. (eds.) Web 2.0 & Semantic Web. AIS, vol. 6, pp. 25–54. Springer, Heidelberg (2010)
12. Kim, H., Breslin, J.G., Decker, S., Choi, J., Kim, H.: Personal knowledge management for knowledge workers using social semantic technologies. Int. J. Intell. Inf. Database Syst. **3** (1), 28–43 (2009)
13. Sauermann, L., Bernardi, A., Dengel, A.: Overview and outlook on the semantic desktop. In: 1st Workshop on the Semantic Desktop at ISWC 2005, vol. 175. CEUR-WS (2005)
14. Drăgan, L., Delbru, R., Groza, T., Handschuh, S., Decker, S.: Linking semantic desktop data to the web of data. In: Aroyo, L., Welty, C., Alani, H., Taylor, J., Bernstein, A., Kagal, L., Noy, N., Blomqvist, E. (eds.) ISWC 2011, Part II. LNCS, vol. 7032, pp. 33–48. Springer, Heidelberg (2011)

15. Bontcheva, K., Cunningham, H., Roberts, I., Tablan, V.: Web-based collaborative corpus annotation: requirements and a framework implementation. In: Witte, R., Cunningham, H., Patrick, J., Beisswanger, E., Buyko, E., Hahn, U., Verspoor, K., Coden, A.R. (eds.) LREC 2010 Workshop on New Challenges for NLP Frameworks, pp. 20–27 (2010)
16. Chamberlain, J., Poesio, M., Kruschwitz, U.: Phrase detectives - a web-based collaborative annotation game. In: International Conference on Semantic Systems (I-Semantics), pp. 42–49. ACM Press, New York, NY (2008)
17. Uren, V., Cimiano, P., Iria, J., Handschuh, S., Vargas-Vera, M., Motta, E., Ciravegna, F.: Semantic annotation for knowledge management: requirements and a survey of the state of the art. J. Web Semant. **4**(1), 14–28 (2006)
18. Corcho, O.: Ontology based document annotation: trends and open research problems. Int. J. Metadata Semant. Ontol. **1**(1), 47–57 (2006)
19. Torres, D., Skaf-Molli, H., Díaz, A., Molli, P.: Supporting personal semantic annotations in P2P semantic wikis. In: Bhowmick, S.S., Küng, J., Wagner, R. (eds.) DEXA 2009. LNCS, vol. 5690, pp. 317–331. Springer, Heidelberg (2009)
20. Bateman, S., Brooks, C., Mccalla, G., Brusilovsky, P.: Applying collaborative tagging to e-learning. In: WWW 2007 Workshop on Tagging and Metadata for Social Information Organization (2007)
21. Noy, N.F., Chugh, A., Alani, H.: The CKC challenge: exploring tools for collaborative knowledge construction. Intell. Syst. **23**(1), 64–68 (2008)
22. Ardissono, L., Bosio, G., Goy, A., Petrone, G.: Context-aware notification management in an integrated collaborative environment. In: UMAP 2009 Workshop on Adaptation and Personalization for Web 2.0, pp. 23–39. CEUR (2010)
23. Goy, A., Magro, D., Petrone, M., Rovera, C., Segnan, M.: A semantic framework to enrich collaborative semantic tables with domain knowledge. In: IC3K 2015 - Proceedings of the 7th International Joint Conference on Knowledge Discovery, Knowledge Engineering and Knowledge Management, KMIS, vol. 3, pp. 371–381. SciTePress (2015)
24. Goy, A., Magro, D., Petrone, G., Segnan, M.: Semantic representation of information objects for digital resources management. Intelligenza Artificiale **8**(2), 145–161 (2014)
25. Goy, A., Magro, D., Petrone, G., Picardi, C., Segnan, M.: Ontology-driven collaborative annotation in shared workspaces. Future Gener. Comput. Syst. **54**, 435–449 (2016). Special Issue on Semantic Technologies for Collaborative Web
26. Magro, D., Goy, A.: A core reference ontology for the customer relationship domain. Appl. Ontol. **7**(1), 1–48 (2012)
27. Borgo, S., Masolo, C.: Foundational choices in DOLCE. In: Staab, S., Studer, R. (eds.) Handbook on Ontologies, 2nd edn, pp. 361–381. Springer, Heidelberg (2009)
28. Gangemi, A., Borgo, S., Catenacci, C., Lehmann, J.: Task taxonomies for knowledge content. Metokis Deliverable D07, pp. 1–102 (2005)
29. Goy, A., Magro, D., Petrone, G., Segnan, M.: Collaborative semantic tables. In: Molli, P., Breslin, J., Vidal, M.E. (eds.) SWCS 2014 - Semantic Web Collaborative Spaces, vol. 1275. CEUR (2014)

Semantic MediaWiki Communities

Wikis and Collaborative Systems for Large Formal Mathematics

Cezary Kaliszyk[1] and Josef Urban[2(✉)]

[1] University of Innsbruck, Innsbruck, Austria
[2] Czech Technical University in Prague, Prague, Czech Republic
josef.urban@gmail.com

Abstract. In the recent years, there have been significant advances in formalization of mathematics, involving a number of large-scale formalization projects. This naturally poses a number of interesting problems concerning how should humans and machines collaborate on such deeply semantic and computer-assisted projects. In this paper we provide an overview of the wikis and web-based systems for such collaboration involving humans and also AI systems over the large corpora of fully formal mathematical knowledge.

1 Introduction: Formal Mathematics and Its Collaborative Aspects

In the last two decades, large corpora of complex mathematical knowledge have been encoded in a form that is fully understandable to computers [8,15,16, 19,33,41]. This means that the mathematical definitions, theorems, proofs and theories are explained and formally encoded in complete detail, allowing the computers to *fully understand the semantics* of such complicated objects. While in domains that deal with the real world rather than with the abstract world researchers might discuss what *fully formal encoding* exactly means, in computer mathematics there is one undisputed definition. A fully formal encoding is an encoding that ultimately allows computers to verify to the smallest detail the correctness of each step of the proofs written in the given logical formalism. The process of writing such computer-understandable and verifiable theorems, definitions, proofs and theories is called *Formalization of Mathematics* and more generally *Interactive Theorem Proving* (ITP).

The ITP field has a long history dating back to 1960s [21], being officially founded in 1967 by the mathematician N.G. de Bruijn, with his work on the Automath system [13]. The development of a number of approaches (LCF, NQTHM, Mizar) followed in the 1970-s, resulting in a number of today's established ITP systems such as HOL (Light) [20], Isabelle [61], Mizar [17], Coq [11],

C. Kaliszyk—Supported by the Austrian Science Fund (FWF): P26201.
J. Urban—Supported by NWO grant nr. 612.001.208 and ERC Consolidator grant nr. 649043 *AI4REASON*.

P. Molli et al. (Eds.): SWCS 2013/2014, LNCS 9507, pp. 35–52, 2016.
DOI: 10.1007/978-3-319-32667-2_3

ACL2 [32] and PVS [46]. This development was often intertwined with the development of its cousin field of *Automated Theorem Proving* [51] (ATP), where proofs of conjectures are attempted fully automatically, without any human assistance. But unlike ATP systems, the ITP systems allow human-assisted formal theory encoding and proving of theorems that are often beyond the capabilities of the fully automated systems.

Twenty years ago, the anonymous QED Manifesto [1] proposed to start a unified encyclopedic effort formalizing all of today's mathematics. This led to a lot of discussion about suitable logical foundations for such a unified effort, and to two QED workshops in 1995 and 1996 [42]. While no such unified foundations and a top-down managed effort emerged, formal libraries of impressive size have been built since then with ITPs: the Mizar Mathematical Library (MML) contains today over 50000 theorems and the Isabelle/HOL and the Flyspeck project (with the large HOL Light libraries) have some 20000 theorems each. ITP is also becoming an indispensable technology for verifying complex software-assisted proofs, and other complicated (e.g., hardware and software) designs. Other exact sciences, such as economics [39] and physics [2] have started to be fully formally encoded recently.

Recent examples of the large projects in formal mathematics include the completed formal proofs of the Kepler conjecture (Flyspeck) [19], the Odd Order theorem [16], the Four Color theorem [15], and verification of more than a half of the Compendium of Continuous Lattices textbook [8]. Verification of the seL4 kernel [33] and the CompCert compiler [41] show comparable progress in full-scale verification of complicated software. Such projects are often linked to various advances in verification technology, such as strong automation methods [27,28,35] that allow less and less verbose formal proofs, however, ITP still remains very labor-intensive. For example, the Flyspeck project is estimated to take 20 person-years, and the Feit-Thompson project took about twice as much.

This means that such large projects in formal mathematics have also very important collaboration aspects. Sometimes they may be managed in a very tight top-down manner, with the formal terminology, naming conventions and majority of stated lemmas being designed by the project leaders, and only the proofs are written by other team members. This has many advantages in consistency, similar for example to the one-man (or one-committee) design of the "upper ontologies" in common-sense reasoning projects such as SUMO [45] and CYC [49]. Sometimes such total top-down control is however missing, and the projects or libraries grow much more chaotically, similar to the way how Wikipedia grows, i.e., only with a basic set of rules for contributing and a lot of later (possibly collaborative) refactoring of the formal terminology, statements, proofs, and theories.

The combination of all these aspects makes formal mathematics a very interesting domain for several reasons:

1. It allows collaboration of experts from very different mathematical areas, both professionals and students and amateurs. As long as the formal proofs are correct, nobody needs to be interested about who wrote them and how,

and if they understood the defined concepts correctly. All this consistency and proof checking is ensured by the underlying ITP systems. This *power of objectivity* has been noted by John McCarthy who proposed to transfer such mechanisms to other sciences and venues of lives.[1]

2. In a similar way, it allows collaboration with machines, particularly AI and ATP systems. Such systems can use very strong semantic search methods such as automated theorem proving, to assist the humans. Again, the final correctness and consistency of such strong machine assistance is checked by the ITP systems.

3. It has all kinds of relations to less formal domains. A very interesting topic is how to present the formal proofs to humans, and also how to assist humans as much as possible with converting informal mathematics into formal. The latter includes all kinds of specialized editors and authoring environments. Such editors and environments may support various semantic features, can be web-based, allow hyperlinking and inter-linking via Semantic Web techniques to more fuzzy concepts defined in Wikipedia and DBpedia, etc. A related interesting topic is how to connect and re-use the formalizations done in various formal systems, or using various libraries.

The rest of this paper is organized as follows. Section 2 summarizes the main differences between encyclopedic efforts like Wikipedia and the reality of formal mathematics. In Sect. 3 we introduce formal mathematical wikis and their main components, such as formal verification, suitable rendering, versioning, editing, and semantic assistance. Section 4 then discusses the examples of today's formal wikis and their various features and subsystems.

2   Some Obstacles to Wikis for Formal Mathematics

In the informal world, Wikipedia can today be largely considered as *The World Encyclopedia* written in a collaborative wiki style, covering vast number of topics and serving more semantics projects such as DBpedia and Wikidata, which can in turn be considered to be the hub of the Semantic Web. In formal mathematics, there is unfortunately no such unique resource, despite the early QED proposals and efforts. Whether such fragmentation is necessary is not clear, however, the following factors largely contribute to this state.

Disagreement about the logical and mathematical foundations. Mathematicians and logicians have different opinions about what should be the most convenient and the most correct foundations for mathematics, ranging from philosophical topics such as constructive versus classical logic, to more practical topics such as using higher-order logic (HOL) versus set theory versus type theory. There are logical frameworks (such as Isabelle [61] or Twelf [52]) that try to cater for everybody, however those may be viewed as an additional burden by some, and their universality, practicality and foundations may be further questioned. The most widely used systems today commit to just one foundational framework.

---

[1] http://www-formal.stanford.edu/jmc/future/objectivity.html.

Disagreement about the formal language and its mechanisms. This is very similar to the disagreement about the appropriateness of various programming languages for various projects. Analogously to that, the chance of all formalizers ultimately agreeing on some of the current formal languages is quite small.

Disagreement about how formal concepts should be defined and mathematics should be built. Even when within common foundations and when using a common formal language and an ITP system, there are sometimes different approaches to formalizing mathematics. In the Coq system, there are several definitions of real numbers, algebraic hierarchies (monoids, groups, rings, fields, modules, etc.) have been built in several competing ways, etc. In the Mizar Mathematical Library, there are several definitions of graphs, categories, and other structures, and parallel theorems about these parallel concepts. Again, this is quite analogous to the current state in programming, where several programming libraries/projects may address the same task differently, according to the tastes and opinions of its implementors. Unlike in Wikipedia, one usually does not prefer to present the different implementations side-by-side within one library, because that makes the library less focused and compact, possibly leading the library's users to confusion and multiple efforts in the same direction.

All these issues have considerable impact on the kind of collaborative tools that can be useful for formalization of mathematics. Around 2004, inspired by the success of Wikipedia, various proposals started to appear suggesting to speed up the relatively slow formalization efforts by wiki-like methods.[2] Ten years later, it is however still not completely clear what is the right combination of features suitable for fast collaborative formalization. We discuss the main components in the next section. While a number of interesting wiki systems have been produced for less formal mathematics, knowledge engineering, research support, and related domains [9,40,44,62], so far there has been no successful attempt to port such systems to the fully formal setting.

3   Formal Wikis and Their Main Components

Traditionally, formal mathematics used to be constructed locally by a single author, either in an editor or in a toplevel[3] of an interactive theorem prover. The prover immediately checks the steps, using its proof rules and the previous knowledge available in the formal libraries, which are locally installed. After the formal definitions, theorems and proofs are written in a file, the file is typically checked again in batch mode by the prover and included in the locally installed library. Various methods have evolved for making one's results available to others. For a long time this would be just an e-mail to library managers, who would then distribute the new library version using FTP or WWW. In the

---

[2] http://wiki.mizar.org/twiki/bin/view/Mizar/MizarWishlist.

[3] Many proof assistants have been implemented inside programming language interpreters and inherit their toplevels.

last decade, a major step towards fast collaborative development was done by switching to version control systems (VCS), which have become also one of the building blocks of today's formal wikis.

In general, a formal wiki would typically address the following six components to various extent:

Formal verification on the server. This is the defining element of the formalization process. Each text that is submitted to a formal wiki will first be evaluated with respect to a particular *formal verifier* (i.e., a proof assistant, ITP) implementing the logic rules and checking the statements and proofs with respect to the logic and the available formal library. Various result statuses of the formal verification are possible. For example, the formal text might be correctly parsed (all concepts are known and used in a correct way), but not completely proof-checked (some proofs make logical inferences that the proof checker cannot understand, perhaps requiring to refine some steps). Or changes in one part of the library might invalidate another part of the library, putting the library as a whole into an inconsistent state. Depending on the results of the verification, the wiki might either completely reject the text, or accept it with some status, usually also updating its existing library with the text, and/or possibly performing some other actions such as branching the development automatically.

Formal library or libraries available on the server, used to supply mathematical terminology and theorems needed for verification of more advanced texts. This is one of the main ways in which interactive theorem proving differs from completely automated theorem proving, where all the information needed for proving a particular statement is typically included in one file. Such large formal libraries are typically stored efficiently in some pre-compiled fast-loadable format, and their parts are loaded into the interactive provers using various inclusion mechanisms. One of the major issues that need to be addressed in formal wikis is the updating, refactoring and maintenance of such libraries of formal concepts, theorems and proofs, their versioning and suitable rendering.

Versioning. Often one wants to see older versions of various files in the library, track the changes done by various users, or even try to work with a non-default version of the library. For example MediaWiki, Wikipedia's internal wiki engine, uses sequential numbering of files (independent of other files), and for every change, the id of the user that made the change is stored with the file. Viewing of differences and histories is possible, although more limited than in more advanced versioning systems. This simplicity can also be an advantage: Wikipedia does not include branches, tags, etc., and the casual user does not need to understand such features. On the other hand, more advanced version control systems operate with the concept of change sets (or commits), which seem more appropriate when several files need to be simultaneously modified (for example when renaming some function) in order for the library to be in a consistent state. If such library consistency is the main concern which is strictly enforced, the Wikipedia-style file-based version control is insufficient, and such advanced version control systems with simultaneous multi-file commits are a necessity.

(Semantic) Rendering. Many interactive provers include documentation generators that process raw prover input files and generate rendered output. The output of a documentation generator is usually HTML or PDF format, with HTML being of particular interest for wikis. Links between files and items are created, different conceptual elements of the prover input are colored in different color, and sometimes mathematical formulas are rendered in a graphical way. Since we are dealing with a very semantic domain, various sorts of information can be added to make the proofs more understandable by the readers, various dependencies can be explicitly shown, etc. While the authors of formal articles typically know how to read the raw formal texts, the HTML presentation (often including MathML) is one of the main aspects of formal wikis that make formal mathematics accessible to newcomers. One of the main reasons is the typical brevity of the formal language which is often optimized for writing rather than reading, and the ubiquitous overloading of mathematical symbols like +, whose particular meaning in a particular context might be very hard to decipher for a casual reader. For example, the large Mizar Mathematical library defines about 200 different uses of the symbol + (for example addition introduced for natural numbers and addition in a group with addition use this symbol). Suitable HTML rendering can link such symbols to their definitions, or even allow their immediate preview by mechanisms such as mouse-over tooltips. In a similar way, aligning with various less formal resources such as textbooks is possible, and also aligning with Semantic Web resources such as Wikipedia or DBpedia.

Editing. While not strictly necessary, it is very useful to have a server part that allows interactive editing of the formal text in a way that resembles local work. Formal proof editors typically offer many advanced semantic features, such as stepping through a proof with the interactive prover and rendering the resulting proof state. Having some of the most used features as a part of the web-based editor is likely to make immediate browser-based updates of the wiki much more attractive for casual readers.

(Semantic) Assisting Tools. Searching for suitable previously defined terminology and suitable theorems is one of the main activities when formalizing mathematics. The better this process is assisted, the more efficient are the authors of formal articles. As any user of today's web search engines knows, server-based technology allows much more sophisticated search tools that can make use of much more expensive processing steps than the user's machine, and that can also benefit from processing much more data than is available on the user's machine. A typical example would be suggesting hints for a new proof based on machine learning on the large libraries of previous proofs stored in the wikis, or running dozens of automated theorem provers on a particular interactively entered goal, making use of a large ("cloud-based") parallelization on the server.

4   Examples of Formal Wikis and Related Systems

The existing examples and prototypes of formal mathematical wikis (and related systems) can be broadly divided by whom they serve and what is their primary purpose.

4.1   Targeted Formal Wikis: Mizar Wiki and Others

Wikis for a particular formal library or a formalization project are perhaps the most visible examples of a wiki-like formal-proof technology. The main example is the *Mizar wiki* prototype [3,57]. It implements fast parallelized server verification, library update and versioning on the server, and HTML rendering for the Mizar system and its large formal library. The HTML rendering can produce all kinds of additional semantic information such as linking symbols to their definitions, showing of the proof state after each proof step, linking of theorems and concepts to Wikipedia and DBpedia, etc.

The verification and rendering are implemented just as hooks to an advanced distributed version control system (DVCS, `git` in this case). This allows the authors to work with the wiki locally, collaborate with others using just the DVCS features, and replicate and distribute the wiki very easily, just as if it was a standard software project. The distinguished Mizar wiki server receives updates from authorized users (as `git` *pushes*), using authorization systems such as `gitolite`[4], and verifies the updates prior to including them into the main or other branch.

The verification against a particular branch of the large library is done using a smart copy-on-write filesystem such as ZFS or BTRFS, making it possible to keep many versions (possibly for many users) of the large library and compiled files on the server without significant space overhead and repetition of work and making the cloning and creation of new (possibly experimental) versions very cheap and fast. The fast cloning also serves for the verification process, which necessarily first has to update the existing library before the library gets formally re-verified. If the re-verification fails, resulting in the rejection of the library update, the new ZFS or BTRFS clone is simply deleted, and the library lives in its original state. If the particular update of a particular version of the library satisfies the verification policy (and the library is thus in some kind of a consistent state after the update), the library is also re-rendered and made available for browsing and further updates.

Updates can be done either using a web interface or by power-users via git. The updates from git can be used for multifile updates (such as library-scale symbol renaming) that would break the library consistency if done on a file-by-file basis. Rudimentary search tools exist for the Mizar wiki, however most of the advanced semantic search functions are not yet integrated with it, and instead run as separate tools produced for distinguished versions of the Mizar library. Similarly, only basic web editing is implemented.

Similar early prototype wiki has been built for Coq [3], and a number of formalization projects and libraries share today some of these features. For example the Coq Users' Contributions[5] are managed inside a VCS, automatically compiled and rendered, however the versioning and user submission are quite restricted. The Isabelle Archive of Formal Proofs[6] (AFP) allows authors

---

[4] http://gitolite.com/.

[5] http://www.lix.polytechnique.fr/coq/pylons/contribs/index.

[6] http://afp.sourceforge.net/.

to update their entries via a DVCS (mercurial) and also provides automated server-based checking, however the HTML rendering is very limited, and the users typically cannot update other entries. In this sense, the archive resemble an evolving online journal, rather than a massively collaborative wiki or a software project.

4.2   Wikis Focused on Editing Support: ProofWeb

To use a proof assistant, one needs to install some software. In the case of HOL Light one needs OCaml with a particular version of CamlP5 compiled with special flags (different than those used by popular Linux distributions), and HOL Light code itself, possibly with checkpointing software. To use an interface to access the prover, one needs the Emacs mode and one of the supported Emacs versions. The process described above is already complicated, not to mention other operating systems and architectures, or additional desirable patches and libraries, or less commonly used provers. The first web interface to the Coq system, LogiCoq [47], would not provide any support for editing: the whole buffer would be sent with standard HTTP request and refreshes the whole page.

ProofWeb [25] provides a web-interface to various proof assistants, that allows ProofGeneral-style [6] interaction. It implements a client-server architecture with a minimal lightweight client interpreted by the browser, a specialized HTTP server and background HTTP based communication between them. The key element of the architecture is the asynchronous DOM modification technique (sometimes referred to as AJAX - Asynchronous JavaScript and XML or Web application). The client part is stored on the server, and when the user accesses the interface page, it is downloaded by the browser, which is able to interpret it without any installation. The user of the interface, accessing it with the browser, does not need to do anything when a modification is done on the server. Every time the user accesses a prover, the version of the prover that is currently installed on the server is used. The user can access any of the provers installed on the server (ProofWeb supports Coq, Isabelle, Matita, Lego, Plastic, and has a minimal support for HOL Light).

The first wiki for Coq that integrated formal text editing with proof assistant feedback [12] was implemented as an extension of the MediaWiki engine, that used ProofWeb for editing pages and rendered the completed articles with Coqdoc in the viewing mode allowing LaTeX snippets. The convenience of a centralized proof assistant environment made it also appealing for teaching (Fig. 1). By defining special tactics for basic logical rules and proof tree rendering code [31] ProofWeb became a convenient tool for various computer courses and has been to date used in 49 courses at 12 universities [22].

A dedicated web interface has also been developed for Matita [5]. Apart from allowing Unicode input and syntax highlighting, it can better exploit the hypertextual document structure offered by the prover, by providing various annotations and active elements. Similarly the Clide [50] interface is a web-reimplementation of the Isabelle interface. It provides a document-oriented interaction with the prover, and allows collaborative editing of an Isabelle script: the edit operations performed by each user are immediately propagated to all.

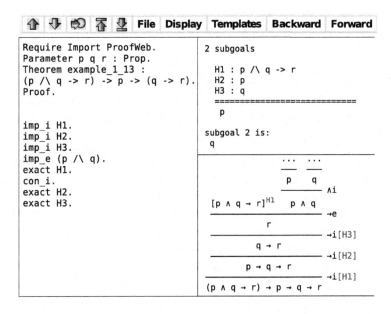

Fig. 1. The ProofWeb interface editing a Coq proof script with a Gentzen-style proof tree.

4.3 Meta Wikis: Agora, Flyspeck Wiki, and Others

While it has so far turned out to be hard to make a one-for-all formalization system and a formal library, the efforts to make a unified interface to different corpora of existing formal mathematics have never stopped.

A recent example is the *Agora wiki* prototype by Tankink [54]. Here, the user combines informal narratives written in the Creole syntax with antiquotations that allow transclusion of formal texts from an arbitrary formal library that has been suitably annotated using the OMDoc ontology developed by Lange [37, 38]. This ontology provides a wide supply of types of mathematical knowledge items, as well as types of *relations* between them, e.g. that a proof proves a theorem. It is a reimplementation of the conceptual model of the OMDoc XML markup language [34] for the purpose of providing semantic Web applications with a vocabulary of structures of mathematical knowledge.

Regardless of the exact details of the formal systems involved, and their output, the annotation process generally yields HTML+RDFa, which uses the OMDoc ontology as a vocabulary. For example, if the formal document contains an HTML rendition of the Binomial Theorem, Agora expects the following result (where the prefix *oo:* has been bound to the URI of the OMDoc ontology[7]):

```
<span typeof="oo:Theorem" about="#BinomialTheorem">...</span>
<span typeof="oo:Proof"><span rel="oo:proves" resource="#BinomialTheorem"/>
...</span>
```

---

[7] http://omdoc.org/ontology#.

The "..." in this listing represent the original HTML rendition of the formal text, possibly including the information that was used to infer the annotations now captured by the RDFa attributes. @about assigns a URI to the annotated resource; here, we use fragment identifiers within the HTML document.

The corresponding RDF annotation has been so far done for the Mizar, Flyspeck and Coq libraries. For instance in Mizar this was done as a part of the XSL transformation that creates HTML from the Mizar's custom semantic XML format [55]. While the OMDoc ontology defines vocabulary that seems suitable also for many Mizar internal proof steps, the current Mizar implementation only annotates the main top-level Mizar items, together with the top-level proofs. Even with this limitation this has already resulted in about 160000 annotations exported from the whole MML. The existing Mizar HTML namespace was re-used for the names of the exported items, such that, for example, the Brouwer Fixed Point Theorem:[8]

```
:: $N Brouwer Fixed Point Theorem
theorem Th14:
  for r being non negative (real number), o being Point of TOP-REAL 2,
      f being continuous Function of Tdisk(o,r), Tdisk(o,r)
  holds f has_a_fixpoint
proof ...
```

gets annotated as[9]

```
<div about="#T14" typeof="oo:Theorem">
  <span rel="owl:sameAs"
        resource="http://dbpedia.org/resource/Brouwer_Fixed_Point_Theorem"/> ...
    <div about="#PF23" typeof="oo:Proof"><span rel="oo:proves" resource="#T14"/> ... </div>
</div>
```

Apart from the appropriate annotations of the theorem and its proof, an additional *owl:sameAs* link is produced to the DBpedia (Wikipedia) "Brouwer_Fixed_Point_Theorem" resource. Such links are produced for all Mizar theorems and concepts for which the author defined a long (typically well-known) name using the Mizar :: $N pragma. Such pragmas provide a way for the users to link the formalizations to Wikipedia (DBpedia, ProofWiki, PlanetMath, etc.), and the links allow data consumers (like Agora) to automatically mesh together different (Mizar, Coq, etc.) formalizations using DBpedia as the common namespace.

A particular instantiation of Agora is the *Flyspeck wiki* prototype [53] used for aligning, cross-linking, and potential further joint refactoring of Hales's informal Flyspeck book [18] about the proof of the Kepler Conjecture, and the corresponding formal Flyspeck development. This alignment takes advantage of the following annotated LATEX form (developed by Hales in his book), which already cross-links informal objects to some of the formal counterparts (formally defined symbols and theorem names living in the formalization):

---

[8] http://mizar.cs.ualberta.ca/~mptp/7.12.02_4.178.1142/html/brouwer.html#T14.

[9] T14 is a *unique internal* Mizar identifier denoting the theorem. Th14 is a (possibly non-unique) *user-level* identifier (e.g., Brouwer or SK300 would result in T14 too).

```
\begin{definition}[polyhedron]\guid{QSRHLXB}
A \newterm{polyhedron} is the
intersection of a finite number of closed half-spaces in
$\ring{R}^n$.
\end{definition}
```

```
\begin{lemma}[Krein--Milman]\guid{MUGGQUF}
Every compact convex set $P\subset\ring{R}^n$ is the convex hull
of its set of extreme points.
\end{lemma}
```

where QSRHLXB and MUGGQUF are the identifiers of the formal definition and theorem. The text contains many further mappings between informal and formal concepts, e.g.:

```
\formaldef{$\op{azim}(x)$}{azim\_fan}
\formaldef{M\"obius contour}{is\_Moebius\_contour}
\formaldef{half space}{closed\_half\_space, open\_half\_space}
```

The wiki relies on MathJaX for rendering the rich informal mathematics, and on custom transformations from LATEX to the Creole wiki syntax. The formal text is automatically included as formal Agora islands, and aligned with the corresponding informal snippets, allowing their simultaneous view.

A related project is the MMT [48] formal logical framework and theory browser, where different formal libraries together with their foundations are *deeply embedded* into the logic of the framework, allowing translations between them and their (at least theoretical) combined use within the framework. The associated HTML theory browser already includes versions of the Mizar [23] and HOL Light libraries [26], however, so far this technology rather serves interested readers than as an user-updatable collaborative wiki.

Another interesting project that combines statistical and semantic methods for aligning the terminologies and theorems in different formal libraries has been started recently [14]. This project shows that quite a lot of such alignment can be achieved fully automatically based on the structure of the library statements. Such tools could complement the manual alignment and annotations used by systems like Agora.

4.4   Server-Based Assisting Tools: HOL(y)Hammer and MizAℝ

In general, collaboration with machines is a very interesting aspect of the fully formal domain. The large formal libraries can be subjected both to various data-mining and machine learning methods that are being developed for less semantic domains such as web search. On the other hand, one can also use very strong semantic search methods such as automated theorem proving (ATP), to assist the humans with finding proofs. Since the final correctness and consistency of such strong machine assistance is checked by the ITP systems (which are today very secure), one can use very efficiently implemented AI/ATP tools as parts of such toolchains, without a risk of introducing an incorrect proof due to implementational errors. The two main examples of such server-based assistance systems are the HOL(y)Hammer system serving the HOL Light users, and the MizAℝ system, serving the Mizar users. Since the two systems are otherwise quite similar, below we only explain their workings for the case of HOL(y)Hammer. The details of MizAℝ can be found in [27,58,59].

HOL(y)Hammer is an online AI/ATP service for formal (computer-under-standable) mathematics encoded in the HOL Light system. The service allows its users to upload or modify and automatically process an arbitrary formal development (project) based on HOL Light, and to attack with automated the-orem provers arbitrary conjectures that use the concepts defined in some of the uploaded projects. For that, the service uses several automated reasoning systems combined with several premise selection methods [36] trained on all the project proofs. The projects that are readily available on the server for such query answering include the recent versions of the Flyspeck, Multivariate Analy-sis and Complex Analysis libraries. The wiki-like features include upload and modification of existing projects, git-based versioning, HTML rendering of the libraries and their optional linking to Wikipedia and DBpedia and production of the OMDoc-based annotations for systems like Agora. The service runs on a 48-CPU server, currently employing in parallel for each task 7 AI/ATP combina-tions and 4 decision procedures that contribute to its overall performance. The current version of the system can prove about 40 % of all Flyspeck toplevel lem-mas fully automatically, thus significantly speeding up the formalization efforts.

The overall problem solving architecture without the updating functions is shown in Fig. 2. The service receives a query (a HOL conjecture to prove, pos-sibly with local assumptions) generated by one of the clients/frontends (Emacs, web interface, HOL session, etc.). The parsed query is processed in parallel by the (time-limited) AI/ATP combinations and the native HOL Light decision pro-cedures (each managed by its forked HOL Light process, and terminated/killed by the master process if not finished within its global time limit). Each of the AI/ATP processes computes a specific feature representation of the query (used for knowledge selection), and sends such features to a specific instance of a premise advisor trained (using the particular feature representation) on pre-vious proofs. Each of the advisors replies with a specific number of premises, which are then translated to a suitable ATP format, and written to a tem-porary file on which a specific ATP is run. The successful ATP result is then (pseudo-)minimized, and handed over to the combination of HOL Light proof-reconstruction procedures. These procedures again run in parallel, and if any of them is successful, the result is sent as a particular tactic application to the frontend. In case a native HOL Light decision procedure finds a proof, the result (again a particular tactic application) can be immediately sent to the frontend.

Since formal projects frequently modify their terminology and theorem names, and updating all the AI/ATP data on the server is expensive, HOL(y)Hammer has introduced *recursive content-based encoding* of the formal theorem and symbol names [56] for all projects and for re-using information across different projects and their versions. This is implemented as follows:

1. The name of every defined symbol is replaced by the content hash (we use MD5) of its variable-normalized definition containing the full types of the variables. This definition already uses content hashes instead of the previously defined symbols. This means that symbol names are no longer relevant in the whole project, neither white space and variable names.

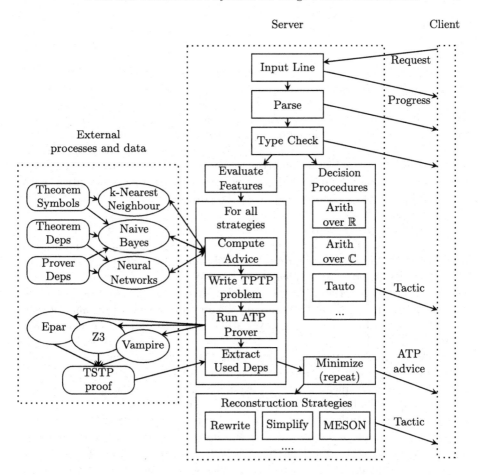

Fig. 2. Overview of the HOL(y)Hammer problem solving functions

2. The name of each theorem is also replaced by the content hash of its (analogously normalized) statement.
3. The proof-dependency data extracted in the content encoding from all projects are copied to a special "common" directory.
4. Whenever a project P is started or modified, we find the intersection of the content-encoded names of the project's theorems with such names that already exist in other projects/versions.
5. For each of such "already known" theorems T in P, we re-use for the AI/ATP systems its "already known" proofs D that are *compatible* with P's proof graph. This means, that the names of the proof dependencies of T in D must also exist in P (i.e., these theorems have been also proved in P, modulo the content-naming), and that these theorems precede T in P in its chronological order of library development (otherwise we might get cyclic training data for P).

This procedure very significantly improves the live updates of the projects, because typically over 90 % of the library's new version remains the same after these normalization steps [29].

5 Conclusions

The world of formal mathematical wikis is still very young, and it is even hard to say that there is a single most advanced system or that a particular wiki-like system has been largely adopted by the formal community. In this article we have therefore tried to summarize the topics that have emerged recently, when the venerable fifty-year old field of formalization of mathematics, with its venerable history of efforts such as QED, got interested in the collaborative wiki-like technologies.

Some of the pieces of the technology have already made it to the formal mainstream. Examples are advanced version control systems that have additional verification and rendering features. Other wiki-related techniques, such as web-based editing supported by verification in ProofWeb, or strong semantic advisors such as HOL(y)Hammer and MizAR have also already found some adopters. The various meta-wiki ideas and ideas linking the formal libraries with semantic web and informal mathematics are however still waiting to be fully developed and utilized.

There are many interesting topics that this brief overview has not dealt with. For example, change propagation is an important research topic and fine-grained dependency tracking can very significantly improve the speed of consistency checking of library updates [4]. Strong semantic assistance can include semi-automated translations between the informal and formal mathematics, with the outlook of gradually training smarter and smarter natural-language-understanding tools for mathematics on such aligned corpora [30]. And very interesting topics arise with future – hopefully wiki-supported – formalizations of other exact sciences such as physics. There, the many possible models of the (never completely known) real world need to somehow formally co-exist, while still being linked to accessible informal explanations about their relations to the underlying reality that they try to capture.

Perhaps the biggest challenge of formal mathematical wikis thus today seem to be the varied and evolving treatments of formality. While fully formal and computer-verified mathematics has made great steps forward in the last decade, there is still no agreement on the ultimate formalism. This includes the particular steps and stages that should link informal and formal knowledge and smoothly proceed in a wiki-like way from one to the other. On the other hand, formal mathematics and the strong semantic tools that have become available for its assistance are already clearly showing how disruptive for our way of doing science could be frameworks that achieve very smooth transition from informal (human-understandable) to formal (machine-understandable) scientific knowledge.

Acknowledgments. The following colleagues have collaborated with us on various aspects of formal wikis and on a number of formal wiki-related systems mentioned in this paper: Mark Adams, Jesse Alama, Grzegorz Bancerek, Kasper Brink, Johan Commelin, Pierre Corbineau, Thibault Gauthier, Herman Geuvers, Mihnea Iancu, James McKinna, Michael Kohlhase, Christoph Lange, Lionel Mamane, Florian Rabe, Piotr Rudnicki, Geoff Sutcliffe, Carst Tankink, Jiri Vyskocil and Freek Wiedijk. Thanks to the anonymous SWCS referees for their valuable comments.

References

1. The QED Manifesto. In: Bundy, A. (ed.) CADE 1994. LNCS, vol. 814, pp. 238–251. Springer, Heidelberg (1994)
2. Afshar, S.K., Siddique, U., Mahmoud, M.Y., Aravantinos, V., Seddiki, O., Hasan, O., Tahar, S.: Formal analysis of optical systems. Math. Comput. Sci. **8**(1), 39–70 (2014)
3. Alama, J., Brink, K., Mamane, L., Urban, J.: Large formal wikis: issues and solutions. In: Davenport, J.H., Farmer, W.M., Urban, J., Rabe, F. (eds.) MKM 2011 and Calculemus 2011. LNCS, vol. 6824, pp. 133–148. Springer, Heidelberg (2011)
4. Alama, J., Mamane, L., Urban, J.: Dependencies in formal mathematics: applications and extraction for Coq and Mizar. In: Jeuring et al. [24], pp. 1–16
5. Asperti, A., Ricciotti, W.: A web interface for Matita. In: Jeuring et al. [24], pp. 417–421
6. Aspinall, D.: Proof general: a generic tool for proof development. In: Graf, S., Schwartzbach, M. (eds.) TACAS 2000. LNCS, vol. 1785, pp. 38–43. Springer, Heidelberg (2000)
7. Autexier, S., Calmet, J., Delahaye, D., Ion, P.D.F., Rideau, L., Rioboo, R., Sexton, A.P. (eds.): AISC/Calculemus/MKM 2010. LNCS, vol. 6167. Springer, Heidelberg (2010)
8. Bancerek, G., Rudnicki, P.: A Compendium of Continuous Lattices in MIZAR. J. Autom. Reasoning **29**(3–4), 189–224 (2002)
9. Baumeister, J., Reutelshoefer, J., Puppe, F.: KnowWE: a semantic Wiki for knowledge engineering. Appl. Intell. **35**(3), 323–344 (2011)
10. Carette, J., Aspinall, D., Lange, C., Sojka, P., Windsteiger, W. (eds.): Intelligent Computer Mathematics. LNCS, vol. 7961. Springer, Heidelberg (2013)
11. The Coq Proof Assistant. http://coq.inria.fr
12. Corbineau, P., Kaliszyk, C.: Cooperative repositories for formal proofs. In: Kauers, M., Kerber, M., Miner, R., Windsteiger, W. (eds.) MKM/CALCULEMUS 2007. LNCS (LNAI), vol. 4573, pp. 221–234. Springer, Heidelberg (2007)
13. de Bruijn, N.G.: The mathematical language AUTOMATH, its usage, and some of its extensions. In: Laudet, M., Lacombe, D., Nolin, L., Schützenberger, M. (eds.) Symposium on Automatic Demonstration. LNM, vol. 125, pp. 29–61. Springer, Heidelberg (1968)
14. Gauthier, T., Kaliszyk, C.: Matching concepts across HOL libraries. In: Watt et al. [60], pp. 267–281
15. Gonthier, G.: The four colour theorem: engineering of a formal proof. In: Kapur, D. (ed.) ASCM 2007. LNCS (LNAI), vol. 5081, p. 333. Springer, Heidelberg (2008)
16. Gonthier, G., et al.: A machine-checked proof of the Odd Order Theorem. In: Blazy, S., Paulin-Mohring, C., Pichardie, D. (eds.) ITP 2013. LNCS, vol. 7998, pp. 163–179. Springer, Heidelberg (2013)
17. Grabowski, A., Korniłowicz, A., Naumowicz, A.: Mizar in a nutshell. J. Formalized Reasoning **3**(2), 153–245 (2010)

18. Hales, T.: Dense Sphere Packings: A Blueprint for Formal Proofs. LMS, vol. 400. Cambridge University Press, Cambridge (2012)
19. Hales, T.C., Adams, M., Bauer, G., Dang, D.T., Harrison, J., Hoang, t.L., Kaliszyk, C., Magron, V., McLaughlin, S., Nguyen, T.T., Nguyen, T.Q., Nipkow, T., Obua, S., Pleso, J., Rute, J., Solovyev, A., Ta, A.H.T., Tran, T.N., Trieu, D.T., Urban, J., Vu, K.K., Zumkeller, R.: A formal proof of the Kepler conjecture (2015). CoRR, abs/1501.02155
20. Harrison, J.: HOL Light: a tutorial introduction. In: Srivas, M., Camilleri, A. (eds.) FMCAD 1996. LNCS, vol. 1166, pp. 265–269. Springer, Heidelberg (1996)
21. Harrison, J., Urban, J., Wiedijk, F.: History of interactive theorem proving. In: Siekmann, J.H. (ed.) Computational Logic. Handbook of the History of Logic, vol. 9, pp. 135–214. North-Holland, Amsterdam (2014)
22. Hendriks, M., Kaliszyk, C., van Raamsdonk, F., Wiedijk, F.: Teaching logic using a state-of-the-art proof assistant. Acta Didactica Napocensia 3(2), 35–48 (2010)
23. Iancu, M., Kohlhase, M., Rabe, F., Urban, J.: The Mizar mathematical library in OMDoc: Translation and applications. J. Autom. Reasoning 50(2), 191–202 (2013)
24. Jeuring, J., Campbell, J.A., Carette, J., Dos Reis, G., Sojka, P., Wenzel, M., Sorge, V. (eds.): Intelligent Computer Mathematics. LNCS, vol. 7362. Springer, Heidelberg (2012)
25. Kaliszyk, C.: Web interfaces for proof assistants. In: Autexier, S., Benzmüller, C. (eds.) Proceedings of the Workshop on User Interfaces for Theorem Provers (UITP 2006), vol. 174, no. 2 of ENTCS, pp. 49–61 (2007)
26. Kaliszyk, C., Rabe, F.: Towards knowledge management for HOL Light. In: Watt et al. [60], pp. 357–372
27. Kaliszyk, C., Urban, U.: MizAR 40 for Mizar 40 (2013). CoRR, abs/1310.2805
28. Kaliszyk, C., Urban, J.: Learning-assisted automated reasoning with Flyspeck. J. Autom. Reasoning 53(2), 173–213 (2014)
29. Kaliszyk, C., Urban, J.: HOL(y)Hammer: online ATP service for HOL Light. Math. Comput. Sci. 9(1), 5–22 (2015)
30. Kaliszyk, C., Urban, J., Vyskocil, J., Geuvers, H.: Developing corpus-based translation methods between informal, formal mathematics: project description. In: Watt et al. [60], pp. 435–439
31. Kaliszyk, C., Wiedijk, F.: Merging procedural and declarative proof. In: Berardi, S., Damiani, F., de'Liguoro, U. (eds.) TYPES 2008. LNCS, vol. 5497, pp. 203–219. Springer, Heidelberg (2009)
32. Kaufmann, M., Moore, J.S.: An ACL2 tutorial. In: Mohamed et al. [43], pp. 17–21
33. Klein, G., Andronick, J., Elphinstone, K., Heiser, G., Cock, D., Derrin, P., Elkaduwe, D., Engelhardt, K., Kolanski, R., Norrish, M., Sewell, T., Tuch, H., Winwood, S.: seL4: formal verification of an operating-system kernel. Commun. ACM 53(6), 107–115 (2010)
34. Kohlhase, M.: OMDoc 2006. LNCS, vol. 4180. Springer, Heidelberg (2006)
35. Kühlwein, D., Blanchette, J.C., Kaliszyk, C., Urban, J.: MaSh: machine learning for Sledgehammer. In: Blazy, S., Paulin-Mohring, C., Pichardie, D. (eds.) ITP 2013. LNCS, vol. 7998, pp. 35–50. Springer, Heidelberg (2013)
36. Kühlwein, D., van Laarhoven, T., Tsivtsivadze, E., Urban, J., Heskes, T.: Overview and evaluation of premise selection techniques for large theory mathematics. In: Gramlich, B., Miller, D., Sattler, U. (eds.) IJCAR 2012. LNCS, vol. 7364, pp. 378–392. Springer, Heidelberg (2012)
37. Lange, C.: OMDoc ontology (2011). http://kwarc.info/projects/docOnto/omdoc.html

38. Lange, C.: Ontologies and languages for representing mathematical knowledge on the semantic web. Semantic Web **4**(2), 119–158 (2013)
39. Lange, C., Rowat, C., Kerber, M.: The ForMaRE project - formal mathematical reasoning in economics. In: Carette et al. [10], pp. 330–334
40. Lange, C., Urban, J. (eds.): Proceedings of the ITP Workshop on Mathematical Wikis (MathWikis), no. 767 in CEUR Workshop Proceedings, Aachen (2011)
41. Leroy, X.: Formal verification of a realistic compiler. Commun. ACM **52**(7), 107–115 (2009)
42. Matuszewski, R. (ed.): The QED Workshop II, Warsaw University Technical report No. L/1/95 (1995)
43. Mohamed, O.A., Muñoz, C., Tahar, S. (eds.): Theorem Proving in Higher Order Logics. LNCS, vol. 5170. Springer, Heidelberg (2008)
44. Nalepa, G.J.: Collective knowledge engineering with semantic wikis. J. UCS **16**(7), 1006–1023 (2010)
45. Niles, I., Pease, A.: Towards a standard upper ontology. In: FOIS, pp. 2–9 (2001)
46. Owre, S., Shankar, N.: A brief overview of PVS. In: Mohamed et al. [43], pp. 22–27
47. Pottier, L.: LogiCoq (1999). URL: http://wims.unice.fr/wims/wims.cgi? module=U3/logic/logicoq
48. Rabe, F.: The MMT API: a generic MKM system. In: Carette et al. [10], pp. 339–343
49. Ramachandran, D., Reagan, P., Goolsbey, K.: First-orderized ResearchCyc: expressiveness and efficiency in a common sense knowledge base. In: Shvaiko P. (ed.) Proceedings of the Workshop on Contexts and Ontologies: Theory, Practice and Applications (2005)
50. Ring, M., Lüth, C.: Collaborative interactive theorem proving with clide. In: Klein, G., Gamboa, R. (eds.) ITP 2014. LNCS, vol. 8558, pp. 467–482. Springer, Heidelberg (2014)
51. Robinson, J.A., Voronkov, A. (eds.): Handbook of Automated Reasoning (in 2 volumes). Elsevier and MIT Press, New York (2001)
52. Schürmann, C.: The Twelf proof assistant. In: Berghofer, S., Nipkow, T., Urban, C., Wenzel, M. (eds.) TPHOLs 2009. LNCS, vol. 5674, pp. 79–83. Springer, Heidelberg (2009)
53. Tankink, C., Kaliszyk, C., Urban, J., Geuvers, H.: Formal mathematics on display: a wiki for Flyspeck. In: Carette et al. [10], pp. 152–167
54. Tankink, C., Lange, C., Urban, J.: Point-and-write - documenting formal mathematics by reference. In: Jeuring et al. [24], pp. 169–185
55. Urban, J.: XML-izing Mizar: making semantic processing and presentation of MML easy. In: Kohlhase, M. (ed.) MKM 2005. LNCS (LNAI), vol. 3863, pp. 346–360. Springer, Heidelberg (2006)
56. Urban, J.: Content-based encoding of mathematical and code libraries. In: Lange, C., Urban, J. (eds.) Proceedings of the ITP Workshop on Mathematical Wikis (MathWikis), no. 767 in CEUR Workshop Proceedings, pp. 49–53, Aachen (2011)
57. Urban, J., Alama, J., Rudnicki, P., Herman Geuvers, A.: Wiki for Mizar: Motivation, considerations, and initial prototype. In: Autexier et al. [7], pp. 455–469
58. Urban, J., Rudnicki, P., Sutcliffe, G.: ATP and presentation service for Mizar formalizations. J. Autom. Reasoning **50**, 229–241 (2013)
59. Urban, J., Sutcliffe, G.: Automated reasoning and presentation support for formalizing mathematics in Mizar. In: Autexier et al. [7], pp. 132–146
60. Watt, S.M., Davenport, J.H., Sexton, A.P., Sojka, P., Urban, J. (eds.): CICM 2014. LNCS, vol. 8543. Springer, Heidelberg (2014)

61. Wenzel, M., Paulson, L.C., Nipkow, T.: The Isabelle framework. In: Mohamed et al. [43], pp. 33–38
62. Worden, L.: WorkingWiki: a MediaWiki-based platform for collaborative research. In: Lange and Urban [40], pp. 63–73

From Ontology to Semantic Wiki – Designing Annotation and Browse Interfaces for Given Ontologies

Lloyd Rutledge[(✉)], Thomas Brenninkmeijer, Tim Zwanenberg,
Joop van de Heijning, Alex Mekkering, J.N. Theunissen, and Rik Bos

Faculty of Management, Science and Technology,
Open University of the Netherlands, Heerlen, The Netherlands
Lloyd.Rutledge@ou.nl

Abstract. We describe a mapping from any given ontology to an editable interface specification for semantic wikis. This enables quick start-up of distributed data-sharing systems for given knowledge domains. We implement this approach in *Fresnel Forms*, a Protégé ontology editor plugin. Fresnel Forms processes any ontology into triples conforming to the Fresnel vocabulary for semantic browser displays. This output format also extends Fresnel with specifications for form-based semantic annotation interfaces. A GUI interface allows editing of this style. Finally, Fresnel Forms exports this interface specification to wikis built with MediaWiki and its extensions Semantic MediaWiki and Semantic Forms.

This work demonstrates Fresnel Forms by creating a wiki that replicates Wikipedia infobox displays and directly exports the triples that DBpedia indirectly derives from them. Forms assist valid user entry of this data with features such as autocompletion. This unifies infobox displays with DBpedia triples while adding assistive data entry in a unified collaborative space.

Keywords: Semantic wikis · Annotation interfaces · Semantic browsers · Fresnel · Style sheets · RDFS · OWL

1 Introduction: Toward a Form-Based Semantic Wikipedia

While the Semantic Web and wikis began as separate visions, various projects have been pulling them together in recent years. Their largest integration begins with Wikipedia[1], a widely used website for information entry and display. Its infoboxes[2] were originally used solely for displaying properties and values for given page topics in small tables along the right side of page displays. Then the DBpedia project converted Wikipedia page links and structured content, including that in infoboxes, into a large set of triples [1] that helped seed the Linked Data Cloud, and has long served as one of its core sources for interlinking concepts. Wikipedia and the DBpedia project have shown how wikis can enable people to enter large amounts of linked data that then gets

---

[1] http://www.wikipedia.org/.
[2] http://en.wikipedia.org/wiki/Help:Infobox.

© Springer International Publishing Switzerland 2016
P. Molli et al. (Eds.): SWCS 2013/2014, LNCS 9507, pp. 53–72, 2016.
DOI: 10.1007/978-3-319-32667-2_4

widely used on the Semantic Web. Wikipedia page infobox content is and remains typed in manually as wiki template code. In addition, its export as semantic data is designed and performed by a third party at a later phase.

The next step in bridging the Semantic Web-wiki gap came in the form of semantic wikis, which add data input and processing to the originally document-oriented wikis. The primary semantic wiki tool is Semantic MediaWiki [16]. Semantic MediaWiki is an extension of the wiki system MediaWiki, which is a widely available and extendable system, and on which Wikipedia runs. It supports making systems that have some equivalents of Semantic Web functionality, such as data annotations and queries on them. In addition, Semantic MediaWiki wikis can export their data in Semantic Web form, thus enabling the immediate posting of a wiki's data on the Linked Data Cloud.

Semantic Forms, an extension of Semantic MediaWiki, brings data and wikis even closer together with Wikipedia infobox-like table displays that define how data derives from their presented contents [15]. In addition, Semantic Forms provides form-based user entry of the data that these displays present and that the semantic wiki can query for. This tool also provides an interface for helping the user create these tables and forms. Semantics Forms is thus an interface-driven developer's tool.

Fresnel is an ontology for specifying interfaces for browsing Semantic Web data, effectively providing triple-defined stylesheets for interacting with data [23]. Typical Fresnel-generated interfaces resemble Wikipedia infoboxes and Semantic Forms tables.

This paper also examines applying the model-driven development approach to making Semantic Web user interfaces. Tools that follow the model-driven development approach in building data systems have the human developer start with a data model from which the system then generates a data browse-and-entry interface that developers can then modify further. This paper applies model-driven approaches to the Semantic Web and semantic wikis. In particular, model-driven development applies here to the generation of Semantic Forms code from the Semantic Web representation of given ontologies.

With this work's results, one can take an ontology and quickly create from it a wiki that helps users enter conforming data and then posts that data directly on the Semantic Web. After presented related work, this paper discusses the use of Wikipedia infobox style as a desired target for testing this approach. Then the paper describes the general architecture for this approach and its implementation. This implementation includes a mapping from ontologies to Fresnel-based code for specifying semantic wiki interfaces. The subsequent section presents this work's primary contribution: the automatic generation of default Fresnel style for given ontologies with our plugin *Fresnel Forms*[3] for the Semantic Web ontology editor Protégé[4]. What follows is a description of various ways how user can tailor this generated default style to efficiently specify the desired interface.

---

[3] http://is.cs.ou.nl/OWF/index.php5/Fresnel_Forms.

[4] http://protege.stanford.edu/.

2   Related Work: Data System Interface Development as Style

The separation of information from its user interface, which is the broader context of this work's goal, is an important practice in several areas of Computer Science. Cascading Style Sheets (CSS) established "style" as how browsers present World Wide Web documents. Meanwhile, the field of model-driven development developed parallels to CSS in the context of defining interfaces to data systems. For the combination of these types of systems, data on the web, Fresnel provides the specification of interfaces to the Semantic Web. Finally, Semantic Forms builds browse and annotation interfaces to data on semantic wikis. This section presents these various approaches to style for information systems, with the aim of unifying them in the rest of this paper.

2.1   Cascading Style Sheets

Cascading Style Sheets (CSS) define how XML documents should appear on web browsers [4]. They *cascade* in the sense that a specific style sheet can build upon the style defined in a more general style sheet. Appendix D of the CSS specification defines a *default* style sheet for HTML [4]. Thus, in the absence of any CSS for an HTML document, a browser renders the document as if applying this default CSS to it as if it were an XML document. Any CSS defined for an HTML document effectively cascades on this default style. This provides efficiency in defining style for HTML because a web developer only needs to specify the difference from the default style. Display according to the default style occurs in the absence any style code.

HTML is, of course, only one type of XML document set. Several browsers, such as Internet Explorer [13], assume default style sheets for all XML documents. These define how the browser displays an XML document in the absence of CSS, and what foundational style any provided CSS builds upon. We see the default style sheet for XML as a *domain-independent style* sheet and the default style sheet for HTML as the default for a specific domain of XML documents. This work applies these concepts to style for the Semantic Web by considering a default style for all Semantic Web data, default style sheets for ontology-defined domains that cascade on the domain-independent default, and the ability to efficiently adapt both of these defaults for efficient tailoring of style.

2.2   Model-Driven Development

Model-driven development tools handle data models as the starting point and core for developing information systems [18]. One such tool, Cathedron, generates a system interface for data display and input [17]. Cathedron generates default interfaces for given databases. We compare this approach to automatically generating a style sheet for a given domain in applying it to our work here. In addition, Cathedron provides several means of specifying the interface beyond its default. One is sorting attribute lists in both display tables and input forms for given classes. Cathedron also offers overrides for different layers in the system and interface.

2.3   Semantic Browsers and Semantic Style

Semantic browsers display triples for a given subject in a tabular form similar to that of infoboxes. However, semantic browsers typically have one style of table that displays all properties in the same way. Tabulator is one example of a semantic browser [3]. Similarly, Semantic MediaWiki offers a "Browse properties" feature, which shows all properties for a given wiki page in one table. We see such displays as domain-independent style, equivalent to how web browsers render XML documents without CSS.

Fresnel is a Semantic Web ontology for the presentation of data from given Semantic Web ontologies [23]. It effectively provides style sheets in the form of RDF triples. Fresnel encapsulates property assignment displays into *lenses*, which are typically associated with classes. Fresnel lenses are thus analogous with the class-based displays of Cathedron. Also like Cathedron, Fresnel enables sorting of properties rows in a display. Fresnel also provides referencing to CSS resources that define the style for presenting lens components. Such CSS overrides the default style that would apply. This use of CSS is an example of how Fresnel, like Cathedron, provides default style override in different layers of the interface.

The */facet* (pronounced "slash-facet") browser in the ClioPatria framework processes RDF-defined mappings between domain ontologies along with an interface model in order to generate search-and-browse interfaces for those domains [12]. The /facet interface includes autocompletion to assist the user in entering terms that the underlying ontology recognizes. Its makers suggest that further research could replace their specific interface ontology with a broader one. In addition, they name Fresnel as a candidate for doing so. Our work builds on these ideas from /facet by using RDF-defined mappings between ontology and interface, and by enabling cascading above a default mapping for each domain. We also use Fresnel as the target interface ontology to map to.

These systems and the approach this paper presents fall within established *categories* for ontology-enhanced user interfaces [21]. In this categorization, the domain of our approach is the *real world*. But this is any real world domain because our approach generates an interface from any given ontology. The complexity of our interface ontology is *medium* because it uses Fresnel and processes medium-level ontology constructs from source ontologies. The presentation form is *lists*. The interaction type is *view and edit* for *refining only*. The usage time is both *design and run time*.

2.4   Wikipedia and DBpedia

The best known wiki is, of course, Wikipedia. Through Wikipedia, DBpedia has become a core source of URI's and triples on the Semantic Web [1]. DBpedia-extracted data often populates other ontologies, including FOAF [7]. The infobox is the main component in Wikipedia that provides these triples. Figure 1 shows an example display for a Wikipedia infobox. Its two-column table structure resembles that of typical semantic browsers: properties are the left, objects and values are on the right, and each row displays a triple with the current page as its object. This is the structure that

Fig. 1. Wikipedia's page display for Tim Berners-Lee, with its infobox on the right (http://en. wikipedia.org/wiki/Tim_Berners-Lee).

DBpedia applies when generating DBpedia triples from Wikipedia infoboxes. These infoboxes also form the basis for making the DBpedia ontology that these triples populate.

We observe here that this Wikipedia-DBpedia workflow is the reverse of the typical data system workflow. Wikipedia-DBpedia workflow is, roughly speaking, appearance-data-ontology. That is, Wikipedia users make infoboxes to form displays of information. Then DBpedia takes these infoboxes, generates triples from them row-by-row, and in doing so forms the DBpedia ontology. A more typical data system approach is to start with a data model or ontology, and then craft a user interface for adding and browsing data populating this ontology. This paper explores having Wikipedia's workflow start with the ontology and generate the user interface from it.

Master's Thesis research at our faculty has used the sorting of properties in Wikipedia infoboxes as a target truth set for various property sorting algorithms [20]. This research shows that a simple ontology-based heuristic for property sorting out-performs random and alphabetic sorting as well as various text-analysis bases algo-rithms. We apply this use of Wikipedia as a truth set here in this work by examining property order and other aspects of infoboxes as target semantic browsing display style.

2.5   Semantic Wikis

Semantic MediaWiki is supported well and remains actively developed. It introduces code syntax for inclusion in wiki page code to annotate the concept a given page represents [16]. Data queries and formatted reports provide access to and presentation of this data. The tool is developed in a Semantic Web research context, but most of its data processing is not strictly as Semantic Web data. However, it does offer an RDF export of the data it manages, including in the form of a directly connected SPARQL endpoint.

Semantic Forms extends Semantic MediaWiki with infobox-like data displays and input forms for populating them [15]. Like Cathedron and Fresnel, Semantic Forms provides grouping and sorting of properties in tabular displays. The fields and parameters that Semantic Forms provides for defining these displays and input forms imply an interface model. Like/facet, Semantic Forms provides autocompletion for annotation entered with its forms. We explore here how to combine Fresnel with a Semantic Web definition of this implied interface model. Although Wikipedia uses neither Semantic MediaWiki nor Semantic Forms, it does apply MediaWiki to display data in tables and, via DBpedia, export that data to the Semantic Web.

Semantic Classes is a proposal for the creation of an extension to Semantic Forms offering a unified specification in XML for the various means of entering and presenting data for a given category [14]. Its proposed functionality is similar to the technique we present here. A key difference in our approach is the integration of existing web style technologies.

Early examples of semantic wikis include SweetWiki [8], with a focus on folksonomic social tagging, and IkeWiki [25], with an early focus on direct Semantic Web compatibility. *OntoWiki* is alternative to MediaWiki and Semantic MediaWiki for Semantic Web functionality on a wiki interface [11]. It offers in-page data like Semantic MediaWiki and a form-based data entry interface comparable to that of Semantic Forms. While we use MediaWiki for our wiki environment, the work techniques presented here are equally applicable in other wiki systems such as OntoWiki. This paper builds upon these other wiki approaches by generating default Fresnel-defined interfaces for given ontologies, and allowing users to modify this style.

Wikidata is a project by Wikimedia that aims to provide a wiki for entering data in the context of other Wikimedia project such as Wikipedia [26]. Part of Wikidata's goal is to generate Wikipedia infobox displays from Wikidata-generated Semantic Web data. This contrasts with the human-written infobox template calls that Wikipedia currently uses. Thus, Wikidata aims for the same kind of display from the same type of data source that this paper does. This paper's style approach can thus apply to efficient specification of the appearance of Wikidata infoboxes.

2.6   Previous Work

In earlier work, we argued for "smart style" in the presentation of RDF by applying and adapting technologies for presenting XML documents on the World Wide Web [19]. Later, we introduced OWL Wiki Forms (OWF) as courseware for Semantic Web courses [24]. OWF lets students enter ontologies with the wiki and then see what

Table 1. Construct mapping from ontologies via Fresnel to various wiki technologies.

Ontology				Fresnel	MediaWiki and extensions			
Foundation – URIs	Whole					[[EquivalentURI::...]] SMW			
	Name-space	Identifier				Identifier/prefix specs OWF	[[Imported from:::...]] SMW		
		Prefix					Wiki page name prefixes MW		
	owl:Thing			:allProperties	Each gets own template and form for domainless properties OWF			
				:hideProperties				
	owl:Ontology							
	rdfs:isDefinedBy							
	owl:imports				Loaded as part of ontology to process OWF			
	rdfs:seeAlso				Link from page for property or category OWF			
Foundation – Text	Fragment identifier							
	rdfs:label			:label	Pagename MW, Label on form & template OWF			
	skos:prefLabel							
	rdfs:comment				Mouseover on label OWF, content on page OWF			
	xml:lang				Selection of text display from user for label, comment, etc.			
	Delimiters	Default Frensel		Additional content	delimiter= SF			
		Cascading Fresnel			Put tekst: before, after, between, start, end, if empty OWF			
CSS				Psuedo-classes	Link style MW			
				:containerStyle	CSS MW — For template and form tabel SF	Whole table		
				:resourceStyle		Row — Multiple		
				:propertyStyle		Row — Single		
				:labelStyle		Cell — Left		
				:valueStyle		Cell — Right, class= SF for forms		
rdf:	:type				[[Category:]] MW			
	Containers			:member in :showProperties	#arraymap/list SF			
	:Property				Property: SMW			
	:Class			:Lens	Page — Category:... MW, also: [[Category:...]] in template MW			
					Page — Template:... MW (if domain)			
					Page — Form:... SF (if domain)			
				:classLensDomain	[[Has default form:::...]] on category SF			
					Check box to assign classes if not domain OWF			
rdfs: – :domain	Fresnel not automatically generated from ontology			:showProperties	Asssign property in template OWF			
				:hideProperties	Un-			
				:showProperties rdf:List	Sort properties in template OWF			
				:mergeProperties :alternateProperties	Template row queries multiple properties for one display SF			
	Class				autocomplete on category=... SF			
					[[Has default form:::...]] on property SF			
rdfs: – :range	Default w/o :range	owl:ObjectProperty			Page			
		owl:DataProperty						
		owl:AnnotationProperty			String			
	range value is — Literal				[[Has type:::...]] SMW — [[Allows value:::...]] SMW			
		textual[1]						
		:language, date parts[2]						
		numeric[3]			Number			
	xsd:	:gYear						
		:date(Time)			Date			
		:time						
		:Boolean			Boolean			
		:URI						
	Fresnel not automatically generated from ontology		:value	:externalLink	URL	Shows as — Linked URL text MW		
				:image		Shows as — Image MW		
				:uri	String	Shows as — Unlinked URL text SMW		
	Semantic MediaWiki only				Geographic coordinate, Code, Temperature			
owl: – Cardinality	min				#arraymap/list SF	repeated fields SF	mandatory SF	
					by default	by default not	✓	
		=1				×		
	max						by default not	
		> 1					✓	
	min				✓			
	Default					×		

Status:	Implemented	Planned or under development

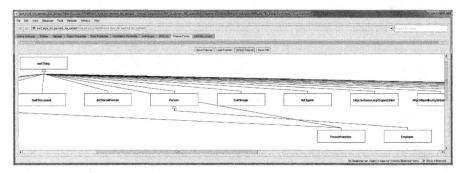

Fig. 2. Screen display of the Fresnel Forms plugin for Protégé. Active ontology is extracted from our DBpedia ontology extract. Current Fresnel is the generated default.

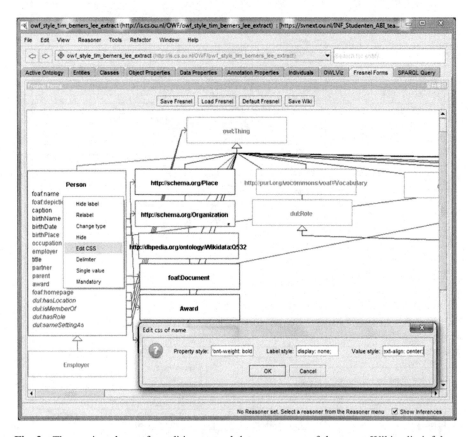

Fig. 3. The session above after editing toward the appearance of the target Wikipedia infobox.

browse and annotations interfaces they derive. We now extend OWF in this paper with the processing of preexisting external ontologies instead of user-entered ones, and by enabling user tailoring with Fresnel style.

3 Technique: Adapting Fresnel Style from Generated Defaults

We develop here a technique for processing Semantic Web ontologies to generate browsing and annotations interfaces. Our technique is to map the components of a given ontology to Fresnel data for its display style, which in turn derives Semantic Forms form pages and template parameter values that define an interface for annotating and browsing instances from that ontology. Constructs in the formats RDF [2], RDFS [5] and OWL [22] serve here as the mapping input. Table 1 shows Fresnel Forms's mapping from ontological constructs to Semantic MediaWiki and Semantic Forms constructs, and the Fresnel constructs bridge them.

3.1 Implementation: A Protégé Plugin

Our implementation of this paper's approach is *Fresnel Forms*, a plugin for the ontology editor Protégé. Fresnel Forms processes OWL ontologies and Fresnel style triples to generate semantic wiki interfaces in the form of MediaWiki XML code that applies constructs from Semantic MediaWiki and Semantic Forms.

The initial input to Fresnel Forms is the active ontology in the current Protégé session. The initial output is the automatic creation of Fresnel code for semantic browsers that works with data from that ontology. During a Protégé session, the user can go to the Fresnel Forms tab and click on its "Generate Default" button. The GUI editor then displays the generated interface. The user can then click on "Save Fresnel" to save this generated default interface as an RDF file of Fresnel triples.

Fresnel Forms can then process this Fresnel code into wiki pages for categories, properties, templates and Semantic Forms form pages. The generated Semantic Forms pages provide assistive form-based data entry. Fresnel Forms replaces Wikipedia's infoboxes with info*rm*boxes, which are MediaWiki templates that display page data on an infobox-like table and define the form with which the user enters that data. The user clicks on "Save Wiki" to start this conversion. A target wiki can then import the wiki code from the XML file that Fresnel Forms saved.

3.2 Generating Default Fresnel Style from the OWL Ontology

The Fresnel Forms approach is to generate a default style for a given ontology in the form of Fresnel triples. This generation of default Fresnel triples can apply in principle to any system that processes Fresnel to set up semantic user interfaces. The Fresnel Forms implementation generates this default Fresnel style and then processes it to create the semantic wiki components that make up the corresponding interface.

In Fresnel, a "lens" is one type of data display, equivalent to an infobox. The core aspects of our generation of default Fresnel style are creating Fresnel lenses for classes that need them and assigning properties to each lens with the fresnel:showProperties property. Fresnel Forms processes property domains to do so. It first makes a lens for each class that is a domain for at least one property. This includes the class owl:Thing, thus also creating an informbox for properties with no explicit domain. Fresnel Forms then assigns properties to the lenses for the classes that are their domains.

Figure 2 shows the Fresnel Forms plugin display after the "Default Fresnel" button has been clicked. The ontology loaded into this Protégé session is an extract we made from the full DBpedia ontology for Wikipedia's Person infobox. We discuss our use of this extract and Wikipedia's infoboxes as a case study in Sect. 4. This GUI display shows a box for each lens, where each lens was generated for a class in the ontology. This display draws lines showing the subclass hierarchy between lenses whose default classes have subclass relationships. At this point, no box shows the properties its lens displays.

Clicking the "Save Wiki" button lets the user save an XML wiki export file. The user can then import this file into a wiki to implement on that wiki the interface this Fresnel code specifies. This conversion processes Fresnel lenses to generate corresponding informbox displays along with their corresponding forms. Its processing of the source ontologies also generates the following wiki components: MediaWiki categories and templates, Semantic MediaWiki property pages and property assignments in other generated wiki pages, and Semantic Forms form pages. Processing property data types determines how their values appear on page displays and how the user enters them on a form. Semantic Forms autocompletion parameters are derived from property ranges.

When the fresnel:showProperties property value is an rdf:list then order of the property URIs in that list is significant. This order becomes the order of those properties in the corresponding informbox on the generated interface. Fresnel Forms's default Fresnel generation of fresnel:showProperties triples uses rdf:list in its values. This order is, however, has no original significance; it is simply the order that the internal API call returns the properties in. When the user changes this order, the rdf:list in the same fresnel:showProperties triples then has this new sequence. This default generation of property order could be improved by applying a sorting heuristic based on ontological structure [20].

Given a generated default style for a given set of ontologies, there are various ways of adapting it for further tailoring semantic wiki interfaces. These means are discussed in the following subsections.

3.3   Editing Fresnel Forms Style with the GUI

In addition to generating default interfaces from given ontologies, the Fresnel Forms Protégé plugin lets the user edit the interface specification with a graphic user interface. Figure 3 shows this GUI during a user editing session on the default interface specification in Fig. 2. Functionalities that this figure shows include:

- *Repositioning* boxes in the GUI display
- *Expanding* boxes to show properties
- *Hiding* lenses and properties
- Hiding or changing property *labels*
- Editing the *CSS* style for property display components
- Setting the *delimiter* displayed between multiple values of a property

The "Save Fresnel" button lets the user save the current interface specification as a local file. Much interface information in this save file is in the form of triples using URIs and structures from the Fresnel ontology. We have expanded the ontology behind this saved triple file to store information beyond the scope of Fresnel. One such expansion saved changes in the GUI display, such as box position and which boxes are expanded.

The upcoming subsections discuss certain aspects of editing Fresnel Forms style in more detail.

3.4   Cascading Ontologies

One means of cascading style with the Fresnel Forms approach is by cascading ontologies. Here, a cascading ontology is simply an ontology that builds upon other ontologies by relating its resources with those of others. This can apply to this paper's example by adding the properties that appear in the Wikipedia person infobox but do not appear in the input ontologies. Making a new ontology with those properties and making their domain be foaf:Person will cause the generated default to include those properties in the generated Fresnel lens for foaf:Person.

3.5   Editing and Cascading Fresnel

This approach cascades ontologies simply by adding new triples that extend the data model. It can also cascade Fresnel style beyond the generated domain-specific default simply because it processes Fresnel triples regardless of their origin. That is, Fresnel Forms processes both the generated default Fresnel triples and any relevant Fresnel triples entered in the triple store by a human user. This is similar to how HTML browsers process referenced CSS stylesheets that website administrators make, which cascade on the browser's default CSS stylesheet for HTML. When generating the wiki interface, a system makes no distinction between the edited cascading Fresnel triples and the generated default Fresnel triples.

Such cascading Fresnel style applies in this paper's example with fresnel: hideProperties triples that block inclusion of certain FOAF properties in the wiki interface. The fresnel:showProperties can also apply to sort the properties that remain on the informbox to have the same order as the target infobox by making its value be an rdf:list ordering those properties.

Another way to override default rendering is to display an image as an image instead of as the URI for that image, which is the default behavior. To achieve this, one can add a Fresnel triple to the Fresnel Forms specification that assigns fresnel:image to the fresnel:value property for formatting foaf:image property values. The Fresnel Forms plugin GUI lets one do this by selecting "Change Type" from the pulldown menu from a second mouse click on the property.

Our Fresnel Forms implementation currently supports only a subset of Fresnel constructs in its querying for Fresnel style for generating wiki interfaces. However, this approach of cascading ontologies as input to generating default Fresnel style that can combine with beyond-default cascading Fresnel style can be applied to making interfaces on any Fresnel implementation. Furthermore, the Fresnel Forms implementation can be used to input the external and cascading ontologies to generate the default Fresnel code than can then be exported for use in other Fresnel implementations.

3.6 Cascading CSS

While Fresnel introduces constructs for defining the general structure of semantic data interfaces, it refers to CSS code for defining the more detailed aspect of page rendering that CSS defines. This approach lets one add Fresnel triples that refer to CSS resources. When processed, such Fresnel-linked CSS will cascade over any CSS that is otherwise linked to the generation of the browser display. Fresnel Forms-generated systems can thus cascade CSS stylesheets via Fresnel.

3.7 Cascading Wiki Templates

There are, of course, aspects of semantic wiki interfaces that neither ontologies nor Fresnel nor CSS can define, and thus that only semantic wiki page code can specify. OWF, the predecessor for Fresnel Forms, allows developers to effective cascade wiki

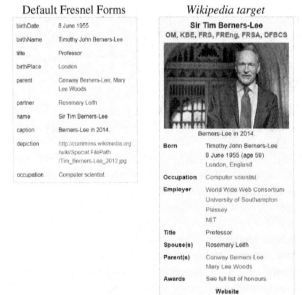

Fig. 4. Informbox display from the default and after editing in Fresnel Forms, compared to original Wikipedia display. All three images are of the same scale.

code on top of the wiki code generated by processing the default style specification and any cascading ontologies, Fresnel and CSS. Each generated informbox template from OWF checks for the existence of a template on the wiki named "InformboxTop" followed by the informbox's name. If a wiki developer had created such a template then it gets transcluded just after the informbox on the rendered wiki page display. An example use case for such cascading wiki templates is the addition of a specialized wiki-generated display on any page for a given type of informbox.

4   Case Study: Wikipedia Infoboxes with Fresnel Forms

The evaluation of the technique presented in this paper has three parts: an implementation, a test input ontology and a target interface. The implementation shows that this paper's technique can be built into a tool. It also provides a system to run tests on. Test input for the system is a widely used ontology. The target output for the system is a selected widely used interface whose data content corresponds with the test ontologies.

This work's evaluation technique is the comparison of an established handcrafted interface for data browsing and entry in a given domain with interfaces automatically generated from Semantic Web ontologies for that domain. Our source for established data browsing interfaces is Wikipedia's infoboxes. The selection criteria applied here for an infobox are that Wikipedia uses it frequently, and that it corresponds with a frequently used ontology. Wikipedia's database report "Templates transcluded on the most pages"[5] shows how frequently each Wikipedia template, including infobox templates, is used. This report shows that the *person* infobox is among the three most frequently used infoboxes.

The online services and prefix.cc analyzes how frequently namespaces are used on the Semantic Web [9]. It can thus indicate an ontology's popularity by the statistics for namespaces it is uses. Of the more than 4000 namespaces the site counts, the most popular namespace for a domain-specific ontology is FOAF[6] [7]. The FOAF ontology has important properties in common with those shown on the infobox for persons. We thus choose the person Wikipedia infobox and the FOAF ontology as representative test sets for testing the feasibility of this paper's approach. Future work can apply additional similarly prioritized infoboxes and corresponding ontologies to broaden this evaluation.

The existing interface serves as the target for what this technique generates in two ways. One is as the interface's general model and form, which the technique is a mapping to in its processing of ontologies. The other is way as an existing example of the interface, which serves as a desired interface for the given knowledge domain. Our technique should be able to generate an interface that is close to this example when processing ontologies from the same knowledge domain. A Wikipedia infobox applies here as the target browsing form. Our technique's aim is thus to generate browsing

---

[5] http://en.wikipedia.org/wiki/Wikipedia:Database_reports/Templates_with_the_most_transclusions.

[6] http://www.foaf-project.org/.

interfaces that are as close as possible to corresponding existing infoboxes. Given this benchmark ontology and interface, and a technique for using them, we show how Fresnel Forms can process the former into the later.

4.1 Default Fresnel Style

The left side of Fig. 4 shows the generated informbox for Tim Berners-Lee from this default Fresnel Forms style. The middle image of Fig. 4 shows how Wikipedia displays the infobox for the same data. One can visually compare the two and see they display the same information in basically the same two-column table format, but the ordering and style of the properties and their values differ.

The primary similarity between these displays lies in the presentation of property values. Object property values are shown as blue links to other wiki pages. In our wiki implementation, wiki pages serve as the resources that URI's in the wiki's data collection refer to. This is typical of how Semantic MediaWiki treats pages as resources to export to RDF as URIs. Data property values are both displays are black and not linked. In addition, default-generated wiki interfaces from Fresnel Forms present values of particular data types appropriates, such as in this case a date.

One difference between the default informbox and the Wikipedia infobox is the ordering of the properties. As the previous section describes, Fresnel Forms does not normalize the default ordering of properties. Another difference is immediately visually obvious: Fresnel Forms cannot automatically detect which URIs reference images that informboxes should display as images. A subtler difference is that Fresnel Forms uses commas by default as delimiters between property values, while this infobox uses newlines. In addition, there are small variations in what data is displayed. The final difference we discuss here is the visual style that CSS defines, such as component size, variations in text size, and general layout.

The next subsection describes how Fresnel Forms can edit the default interface to be closer in appearance to the target style. It also analyses what aspects of the target interface Fresnel Forms cannot implement.

4.2 Edited Fresnel Style

Figure 4 shows how close the resemblance is between the edited Fresnel Forms style on the right and the original Wikipedia infobox for Tim Berners-Lee in the middle. While the default style captures basics, such as which properties to show in each class's lens, and how to display property values, editing brings the visual appearance much closer to the target, although not completely.

Figure 3 shows the Fresnel Forms plugin display while editing the default style to resemble the Wikipedia target infobox display more closely. With this GUI, the user is able to generate Fresnel code that makes the informbox appear more like the corresponding Wikipedia infobox. The user changes the property order by holding a property in a lens box in the GUI with a mouse button and dragging the property up or down in the list. This results in that property's URI having a different position in the rdf:list of the corresponding fresnel:showProperties triple.

Next to the property order in fresnel:showProperties, Fresnel has additional constructs that apply here. We change the text used for naming properties along the left side by using the GUI to set the fresnel:label of that property's Fresnel format to the desired string. In addition, the image appears as an image as the result of clicking "Change type" from the foaf:depiction property's pull-down menu and then selecting "IMAGE". The resulting Fresnel code is a "fresnel:value fresnel:image" triple for the format object for foaf:depiction. We achieve the appearance of clustering the three properties under "Born" by setting the fresnel:label for birthName to "Born" and then assigning "fresnel:label fresnel:none" to birth date and birth place to hide their labels.

Fresnel also has constructs for CSS that apply here. We set the property labels to bold font. The spacing between properties and their values can be adjusted via CSS. To enable the main name and the property website to appear to span both "columns" of the "table", Fresnel Forms defines both with the < div > element along with default CSS that makes them appear of a table. Then we override this default with CSS for that property's display. CSS embedded in Fresnel also sets the size of the image.

We extend Fresnel with new properties that go beyond Fresnel's original scope. Section 3.3 discusses our extension with properties for the positioning of boxes in the GUI's display. We use the prefix "owf:" (OWL Wiki Forms) for the namespace for our extensions to Fresnel. One such extension, "owf:delimiter", sets what the delimiting

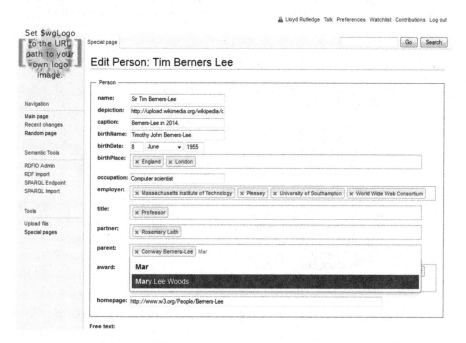

Fig. 5. Example of filling in a Person form from the default interface for our extract from the DBpedia ontology, with autocompletion for the property field "parent".

character is between multiple values of a property in a display. Fresnel Form's wiki export defaults to commas as such delimiters because Semantic MediaWiki and Semantic Forms do as well. In this example, we override this default with a newline value for owf:delimiter for the property lists for employer and parent.

There are several aspects of the Person infobox that we could not achieve with Fresnel Forms. One is the value for "Awards" as a link to a separate page listing the awards as a document rather than as data. We chose not to have a link to a document be a value here. Because the linked document did not structure the awards as data, we didn't try to enable assigning them directly. This is a case where a component of an infobox is intended more for visual document display than for data. Such aspects of infoboxes usually do not directly correspond to a structured, semantic approach.

The calculation and display of the person's current age is readily calculable from information readily acquire from data, but we have not found a graceful what to implement this with Fresnel. The honorific suffixes that the Wikipedia infobox for Tim Berners-Lee displays under his name could be implemented with Fresnel sublenses, but we have not yet implemented them. Lacking Fresnel sublenses, our wiki-based implementation takes the full page name of each honorific. These we display the honorifics instead as values for "awards", which is how they appear in the DBpedia RDF for Tim Berners-Lee. (DBpedia can be out of sync with the current content of Wikipedia infoboxes.) One could annotate each honorific with a suffix and then select the suffix in the sublens for honorifics to get the list of suffixes by Tim Berners-Lee's name. A fuller implementation of Fresnel could thus take the generated display this one step further. This implementation would also involve adding the suffix property to the honorific class, perhaps as a local small extension to the source DBpedia ontology.

```
@prefix dbpedia <http://dbpedia.org/ontology/#> .

dbpedia:birthDate rdf:type owl:DatatypeProperty ;
    rdfs:label "BirthDate" .

wiki:Tim_Berners-2DLee a wiki:Category-3APerson,
        swivt:Subject ;
    rdfs:label "Tim Berners-Lee" ;
    dbpedia:BirthDate "1955-06-08"^^xsd:date ;
    dbpedia:BirthName "Timothy John Berners-Lee"^^xsd:string ;
    dbpedia:BirthPlace wiki:London ;
    dbpedia:Caption "Berners-Lee in 2014."^^xsd:string ;
    dbpedia:Modification_date-23aux 2.457075e+06 ;
    dbpedia:Name "Sir Tim Berners-Lee"^^xsd:string ;
    dbpedia:Occupation wiki:Computer_scientist ;
    dbpedia:Parent wiki:Conway_Berners-2DLee,
        wiki:Mary_Lee_Woods ;
    dbpedia:Partner wiki:Rosemary_Leith ;
    dbpedia:Title "Professor"^^xsd:string .
```

Fig. 6. Extract from RDF export of example Tim Berners-Lee page.

4.3   Form-Based Annotation with an Extension of Fresnel

Fresnel defines interfaces for semantic browsers but not specifically for semantic annotators like Semantic Forms. Our implementation applies some Fresnel style to forms as well as informboxes, such as property orders. We also identify some features that are specific to form-based annotation, and that Semantic Forms implements. We define these features in our extension to Fresnel Forms. Fresnel Forms' wiki conversion then applies the appropriate Semantic Forms parameters to forms generated using these new features. Figure 5 shows the form-based annotation interface that the edited Fresnel Forms style generates on a wiki, with data entered for Tim Berners-Lee.

One aspect of form-based annotation that our extension covers is the data type of the input field. Figure 5 shows birth date being entered as a date. The wiki code generate specifies this with the value "date" for the Semantic Forms parameter "input type". Our extension uses the "owf:datatype" property with data types from XML Schema to store this information.

Autocompletion is another helpful feature in form-based annotation in general and in Semantic Forms in particular. Semantic Forms enables form fields to have autocompletion to given categories. In Fresnel Forms, this means the forms for some properties have autocompletion to given classes. The property owf:autocompleteFromClass specifies to which class a property's value should autocomplete in the form. By default this goes to the class that is the rdfs:range for the property. Figure 5 demonstrates autocompletion from the parent field to members of the Person class.

4.4   RDF Export

A main goal of the wiki in this scenario is to generate data in RDF. Figure 6 shows an extract from the wiki's RDF export from the example Tim Berners-Lee page. The wiki exports the birthDate property as the original URI from the DBpedia ontology by applying the [[Imported from::]] functionality from Semantic MediaWiki. This is the wiki's final step in helping the user enter data that is valid for the ontology that this workflow started with.

4.5   Scaling to All Wikipedia Infoboxes

Until now, this case study has focused on a single infobox from Wikipedia: that for Person. Of course, to make a full contribution to the interface for Wikipedia, this tool should be able to handle all of the other infoboxes as well. This subsection extends the previous discussion of the quality of the implementation of a single infobox to how Fresnel Forms scales to handle the full quantity of Wikipedia.

On one laptop, it takes the Fresnel Forms plugin for Protégé several minutes to generate the display for the default Fresnel style for the whole DBpedia ontology[7]. Saving this to a Fresnel file takes about half an hour. Running Protégé on a specialized

---

[7] http://downloads.dbpedia.org/2014/dbpedia_2014.owl.bz2.

server reduces Fresnel file generation time to about half a minute. Exporting to a wiki XML file on this server takes about 20 min. In terms of timing, applying Fresnel Forms to replacing all Wikipedia infoboxes is feasible, albeit it would have to run as a batch job instead of live.

The numbers that correspond with this timing are the following. The DBpedia ontology file is 2.3 megabytes large when uncompressed. Fresnel Forms generates from this a Fresnel file of 18.9 megabytes containing 760 lenses, each corresponding to an infobox. Fresnel Forms then converts this Fresnel into a wiki XML export file of 126 megabytes. This file imports 5146 wiki pages for the informboxes, forms, categories and properties that define the wiki interface.

The wiki that this XML file generates would then have the default Fresnel style from the DBpedia ontology. The human effort applied to make the Person informbox visually resemble its Wikipedia infobox target would then have to be applied to each of the other lenses as well. The Fresnel and wiki files generated for DBpedia are available from this paper's website.

5 Conclusion: Further with Adapting Default Semantic Style

This paper described applying default and adapted style to generate annotate-and-browse interfaces from given ontologies. Our implementation of this approach, Fresnel Forms, is a plugin for Protégé that is readily installed and made accessible online to distributed users. Fresnel is applied to define the interface style. Adoption and adaptation of the CSS approach enable cascading style definitions on top of such default interfaces. This approach is evaluated by replicating a Wikipedia infobox display. While Wikipedia infoboxes serve here as proven interfaces to aim for, this approach can hypothetically implement Wikipedia infoboxes themselves. This approach enables developers to efficiently make and maintain distributed browse and annotate interfaces for given ontologies.

Acknowledgements. This research is funded by the Faculty of Management, Science and Technology of the Open University in the Netherlands. The implementation of Fresnel Forms is the result of two Bachelor Thesis group projects in the faculty's Informatics Bachelor program: *Protégé-OWL MDD Fresnel Plug-in* [6] and *Protégé Fresnel Forms Plugin* [10]. The Fresnel Forms plugin is available for download online[8].

References

1. Auer, S., Bizer, C., Kobilarov, G., Lehmann, J., Cyganiak, R., Ives, Z.G.: DBpedia: a nucleus for a web of open data. In: Aberer, K., et al. (eds.) ASWC 2007 and ISWC 2007. LNCS, vol. 4825, pp. 722–735. Springer, Heidelberg (2007)
2. Becket, D. (eds.): RDF/XML Syntax Specification (Revised), W3C Recommendation (2004)

---

[8] http://is.cs.ou.nl/OWF/index.php5/Fresnel_Forms.

3. Berners-Lee, T., Chilton, L., Connolly, D., Dhanaraj, R., Hollenbach, J., Lerer, A., Sheets, D.: Tabulator: exploring and analyzing linked data on the semantic web. In: Proceedings of the 3rd International Semantic Web User Interaction Workshop (SWUI 2006) (2006)
4. Bos, B., Çelik, T., Hickson, I., Wium Lie, H.: Cascading style sheets level 2 revision 1 (CSS 2.1) specification. W3C Recommendation (2011)
5. Brickley,D., Guha, R.V. (eds.).: RDF vocabulary description language 1.0: RDF schema, W3C Recommendation (2004)
6. Brenninkmeijer, T., Zwanenberg, T.: Protégé-OWL MDD fresnel plug-in, bachelor's group project thesis, Open University in the Netherlands, August 2014
7. Brickley, D., Miller, L.: (eds.) FOAF Vocabulary Specification 0.98 (2010)
8. Buffa, M., Gandon, F.: SweetWiki: semantic web enabled technologies in Wiki. In: Proceedings of the 2006 Symposium on Wikis (WikiSym 2006), Odense, Denmark, August 2006
9. Digital Enterprise Research Institute (DERI). prefix.cc namespace lookup for RDF developers (2013). http://prefix.cc/. Accessed 03 April 2013
10. van de Heijning, J., Mekkering, A., Theunissen, J.N.: Protégé Fresnel forms plugin, bachelor's group project thesis, Open University in the Netherlands, May 2015
11. Heino, N., Dietzold, S., Martin, M., Auer, S.: Developing semantic web applications with the OntoWiki framework. In: Pellegrini, T., Auer, S., Tochtermann, K., Schaffert, S. (eds.) Networked Knowledge - Networked Media. SCI, vol. 221, pp. 61–77. Springer, Heidelberg (2009)
12. Hildebrand, M., van Ossenbruggen, J.: Configuring semantic web interfaces by data mapping. In: Workshop on Visual Interfaces to the Social and the Semantic Web (VISSW 2009), February 2009
13. Howlett, S., Dunmal, J.: Beyond ASP: XML and XSL-based solutions simplify your data presentation Layer, MSDN Magazine, November 2000
14. Koren, Y.: Semantic classes proposal. Semantic MediaWiki Plus - SMW+ A Semantic Web Enterprise Wiki, Semantic Wiki Discussion Session 9 (2010)
15. Koren, Y.: Semantic Forms (2013). http://www.mediawiki.org/wiki/Extension:Semantic_ Forms. Accessed 03 April 2013
16. Krötzch, M., Vrandecic, D., Völkel, M., Haller, H., Studer, R.: Semantic wikipedia. J. Web Semant. **5**, 251–261 (2007)
17. Mattic B.V., Cathedron Manual for the Preview / Field Test Release, September 2007
18. Mellor, S.J., Clark, T., Futagami, T.: Model-driven development: guest editors' introduction. IEEE Softw. **20**(5), 14–18 (2003)
21. van Ossenbruggen, J., Hardman, L., Rutledge, L.: Combining RDF semantics with XML document transformations. J. Web Eng. Technol. (IJWET) **2**(2/3), 248–263 (2005)
20. Paul, F., Property ranking approaches for semantic web browsers - a review of ontology property ranking algorithms. Masters thesis, Open University in the Netherlands, December 2014
21. Paulheim, H., Probst, F.: Ontology-enhanced user interfaces: a survey. In: Semantic-Enabled Advancements on the Web: Applications Across Industries. IGI Global, 2012, pp. 214–238 February 2012. doi:10.4018/978-1-4666-0185-7.ch010
22. Patel-Schneider, P.F., Hayes, P., Horrocks, I., (eds.) OWL Web ontology language semantics and abstract syntax, W3C Recommendation (2004)

23. Pietriga, E., Bizer, C., Karger, D.R., Lee, R.: Fresnel: a browser-independent presentation vocabulary for RDF. In: Cruz, I., Decker, S., Allemang, D., Preist, C., Schwabe, D., Mika, P., Uschold, M., Aroyo, L.M. (eds.) ISWC 2006. LNCS, vol. 4273, pp. 158–171. Springer, Heidelberg (2006)
24. Rutledge, L., Oostenrijk, R.: Applying and extending semantic Wikis for semantic web courses. In: Proceedings of the 1st Workshop on eLearning Approaches for the Linked Data Age (Linked Learning 2011), Heraklion, Greece, May 2011
25. Schaffert, S.: Ike Wiki: A Semantic Wiki for collaborative knowledge management. In: Proceedings of the 15th IEEE Workshops on Enabling Technologies: Infrastructures for Collaborative Enterprises (WETICE 2006), Manchester, U.K. June 2006
26. Vrandečić, D.: Wiki data: a new platform for collaborative data collection. In: Keynote presentation at Semantic Web Collaborative Spaces Workshop (SWCS 2012), Lyon, France, April 2012

A Semantic MediaWiki-Based Approach for the Collaborative Development of Pedagogically Meaningful Learning Content Annotations

Stefan Zander[1]([✉]), Christian Swertz[2], Elena Verdú[3],
María Jesús Verdú Pérez[4], and Peter Henning[5]

[1] FZI Research Center for Information Technology, 76131 Karlsruhe, Germany
`zander@fzi.de`
[2] Department of Education, University Vienna,
Sensengasse 3a, 1090 Vienna, Austria
`christian.swertz@univie.ac.at`
[3] Universidad Internacional de La Rioja, (UNIR), La Rioja, Spain
`elena.verdu@unir.net`
[4] ETSI Telecomunicación, University of Valladolid,
Campus Miguel Delibes, 47011 Valladolid, Spain
`marver@tel.uva.es`
[5] Karlsruhe University of Applied Sciences, Karlsruhe, Germany
`peter.henning@hs-karlsruhe.de`

Abstract. In this work, we present an approach that allows educational resources to be collaboratively authored and annotated with well-defined pedagogical semantics using Semantic MediaWiki as collaborative knowledge engineering tool. The approach allows for the exposition of pedagogically annotated learning content as Linked Open Data to enable its reuse across e-learning platforms and its adaptability in different educational contexts. We employ Web Didactics as knowledge organization concept and detail its manifestation in a Semantic MediaWiki system using import and mapping declarations. We also show how the inherent pedagogical semantics of Web Didactics can be retained when learning material is exported as RDF data. The advantage of the presented approach lies in addressing the constructivist view on educational models: The different roles involved in the content development process are not forced to adapt to new vocabularies but can continue using the terms and classification systems they are familiar with. Results of the usability test with computer scientists and education researchers are positive with significantly more positive results for computer scientists.

1 Introduction

In order to establish good learning content and to introduce adaptive learning systems to the classroom, the constructivist view on educational models claims for the integration of many different roles (e.g., instructors, instructional

P. Molli et al. (Eds.): SWCS 2013/2014, LNCS 9507, pp. 73–111, 2016.
DOI: 10.1007/978-3-319-32667-2_5

designers, pedagogues, media designers, and students) in the learning content development process [1–3]. As a consequence, it should be simplified for both authors and instructors [4] and support the aspect of collaboration [2].

The latest generation of e-learning solutions aim to address this by emphasizing the aspects of decentralization and inter-institutional collaboration, which leads to an increasing necessity of accessing and utilizing learning content outside specific e-learning platforms [4,5]. The realization of such decoupled and unobstructed access requires—among other things—expressive representation frameworks for both the organization and representation of learning content annotations and is of particular relevance in the educational sector [6]. Collaboratively created semantic vocabularies help in improving access for learners and instructors and facilitate the exchange of learning material across different platforms as well as its reuse in different contexts and for different purposes through pedagogically meaningful semantics [4]. Although the benefits that even simple annotation systems offer to the development process of learning content are broadly acknowledged in the e-learning domain (cf. [2,5,6]), the process of integrating lightweight annotation systems such as Semantic MediaWikis into educational systems and learning content generation processes has only recently begun [5,7]. Exacerbating factors are the difficulties in designing and using ontologies as formalisms for representing annotation semantics together with the high engineering experience required by ontology engineering tools that domain experts such as instructors, instructional designers, and content developers usually do not have. Related studies (e.g. [6,8]) also indicate the lack of available domain ontologies for several subjects together with the lack of standards, tools, and design methodologies.

While Semantic MediaWikis do not seem to be the most obvious candidates for the management of e-learning content, they have been suggested as appropriate tools for this task since they address some problems that exist with common learning management systems (cf. [5,7]):

(i) They adopt the Wiki-based authoring style for the creation of lightweight ontologies.
(ii) Semantic MediaWikis are conducive to reaching a shared agreement about the relevant entities in a universe of discourse—an aspect that is of central importance for the acceptance and broad usage of a domain ontology.
(iii) They help in making learning content available for the Web of Data[1] and hence contribute to the recently emerging trend of *educational Linked Open Data* (see [5]).
(iv) Semantic MediaWikis offer a version control system.
(v) They do not only support the management of content within courses, but the creation of large common content repositories.
(vi) Semantic MediaWiki offers RDF support.

As a consequence, Semantic MediaWikis seem to be promising candidates to manage collaboratively maintained content repositories. To support and facilitate

---

[1] cf. http://www.w3.org/2013/data/.

this trend, we present an approach that allows learning content to be collaboratively authored and annotated with well-defined pedagogical semantics using Semantic MediaWiki as collaborative knowledge engineering tool.

1.1 Research Questions and Contributions

With our approach, we show that we can overcome the limited expressivity of Semantic MediaWikis knowledge representation framework by importing the *Pedagogical Ontology (PO)* and the *Semantic Learning Object Model (SLOM)* created in the INTUITEL project[2] in order to create rich pedagogically meaningful annotations that can be processed by INTUITEL-enabled Learning Management Systems (LMS) and are conducive to their utilization in the educational Web of Data [5] with an acceptable usability. More specifically, this work addresses the following research questions:

(i) *How can the rich pedagogical semantics defined in the Pedagogical Ontology and the Semantic Learning Object Model be made available in a Semantic MediaWiki system for collaborative content authoring while maintaining their formal, model-theoretic semantics?*

(ii) *Can pedagogically enhanced Semantic Media Wikis support the arrangement of content for heterogeneous learning sequences in online learning processes?*

(iii) *Do teachers accept the usability of Semantic Media Wikis as a tool for creating multi-sequenced content online?*

In answering that questions, we show how the concept of Web Didactics and its manifestation in the Pedagogical Ontology and the Semantic Learning Object Model can be integrated into a Semantic MediaWiki system using import and mapping declarations. We demonstrate how the rich pedagogical semantics can be retained, although the underlying knowledge representation frameworks are defined on description logics with differing expressivity (see Sect. 4.2). The impact of the presented approach is as follows:

(i) We facilitate the reuse of learning material through its annotation with pedagogically meaningful semantic data defined in the Pedagogical Ontology while minimizing the necessity to use external tools or full-fledged ontology editors such as Protégé [9] to create such annotations.

(ii) The presented approach builds on standardized semantic Web technologies and allows learning content annotations to be exported as Linked Data. This enables learning content to be linked to related content and reused outside specific LMSs and in different contexts.

(iii) It does not reduce or negatively impact the efficiency of existing Semantic MediaWiki-based authoring processes (cf. [4,10]).

(iv) We ensure collaborative authoring since the production of distance learning material requires the collaboration of people with different skills (pedagogues, computer scientists, graphic designer etc.).

---

[2] http://www.intuitel.de/.

As a consequence, course instructors are not forced to learn or adapt to new annotation vocabularies in order to create pedagogically meaningful annotations. They can rather continue using the vocabulary terms and classification systems they are familiar with. The presented approach does also not require an interruption or reconfiguration of content creation processes as appropriate semantics are added during authoring time. Our work builds on the standard Semantic MediaWiki system and can be combined with related approaches, e.g., to add offline editing support or multi-synchronous work mode [11] (see also Sect. 3).

This document is structured as follows: Sect. 2 introduces the Web Didactics as a knowledge organization concept that is focused on the collaborative production and representation of learning content. The manifestation of the Web Didactics in the Pedagogical Ontology and the Semantic Learning Object Model (SLOM) is described, followed by a brief introduction to Semantic MediaWiki as collaborative ontology engineering tool. The unique features of the present approach as well as its differentiation to related works are discussed in Sect. 3. Based on a description of the main characteristics and features of the PO and SLOM, we elaborate in Sect. 4 how the particular semantics of the Pedagogical Ontology and SLOM can be reflected in the model-theoretic semantics of a Semantic MediaWiki system. In doing so, we discuss different aspects related to the exposition of learning content and its pedagogical semantics as Linked Open Data using a Semantic MediaWiki system. In Sect. 5, we validate the presented approach and discuss the extent to which the formal semantics of the PO and SLOM can be preserved in the knowledge representation formalism upon which SMW is built. We also present its general applicability in a real-world use case in which a university lecture about network design is exposed as Linked Data together with its collaboratively created pedagogical semantics. In Sect. 6 results from an usability test where content was created and annotated with the INTU-ITEL PO and SLOM by teachers from departments of computer technology and departments of education are reported. Section 7 summarizes limitations of the presented approach and outlines open issues to direct future work on the given topic. A final verdict is given in Sect. 8.

2   Background

This section provides background knowledge about the concepts, tools, and models that are fundamental for the presented approach. We first introduce the concept of *Web Didactics* and explain its main characteristics followed by an overview of the *Pedagogical Ontology (PO)*—a machine-processable representation of selected aspects of Web Didactics. The Pedagogical Ontology together with *the Semantic Learning Object Model (SLOM)*, which is described subsequently, serves as representation basis for the collaborative eLearning content modeling approach presented in this work. We use a *Semantic MediaWiki (SMW)* as technical basis of our approach and provide an overview of its knowledge representation formalism in the last part of this section, as it serves as collaboration platform for the authoring and annotation of learning content.

2.1   Web Didactics

Individualization of learning processes is an old and well established cultural practice. Skimming through books is one simple example for this cultural practices. In the book culture, this is supported by tools like tables of contents, indexes, page numbers, etc. With computer technology, it is possible to support individual learning not only by static, but also by dynamic tools. Due to the qualities of digital electronic universal Turing machines, dynamic tools to support individualized learning require the expression of the pedagogical meaning of content in a machine readable format [12]. Thus the first purpose of the Web Didactics is to support the expression of the pedagogical meaning of content by offering a metadata vocabulary.

While individualization of learning processes is a well known practice, this is not the case for collaborative authorship of content. To support collaborative content production for individualized online learning requires a classification system which has to be suitable for multiple curricula and pedagogies. Thus the second purpose of the Web Didactics is to offer a classification system that supports collaborative knowledge production for multiple curricula and pedagogies [13]. With this approach, the static classification system can be dynamically transferred into the learning time by taking learner behaviour into account. The classification system is thus a collaborative space for the creation of semantically enriched learning material that can be turned into dynamically calculated recommendations and feedback in the learning process.

The model behind the classification system is based on the distinction of knowledge representation in space and knowledge communication in time [14]. Knowledge representations in space are classified by a metadata system and a metadata vocabulary. Knowledge representation in time is represented by learning pathways. The purpose of the ontology developed by Meder was to create a metadata system and a vocabulary that are suitable to express every teaching and learning concept that was developed in the history of teaching and learning. During runtime, this ontology is combined with a learner ontology that describes the behaviour of an individual learner [15].

Formal knowledge representation in space depends on granularity. While the structure of output devices has to be considered for granularity in the context of teaching and learning, granularity of content still can only be estimated. It is thus a heuristic value. We have suggested, to take the properties of the computer screen as an obviously rough orientation to estimate granularity. In the context of computer technology, we consider the amount of content that can be perceived in 5 to 10 min as the smallest sensible section of knowledge. This we define as a knowledge object (KO). KOs are the first line of the classification of learning objects we suggest. Every knowledge object is described by a set of metadata. The set of metadata includes the knowledge domain of the content, the topic that is covered, author, license, production time and date, level, if it's suitable for blind, deaf or dumb people, minimum screen resolution and file size, age of the targeted audience, language, and media type and knowledge type. The selection criterion for these metadata is applicability. We tested this by describing algorithms about how to apply those criteria in the learning process. If, for example, the learner tells the system that he his blind, the system will try to avoid the recommendation of content that is not suitable for blind people. Obviously, this depends on the availability of appropriate content.

While most of these meta data are self explanatory and the vocabulary is fairly simple, this is not the case for the knowledge types. As knowledge types the Web Didactic offers a vocabulary that distinguishes receptive, interactive and cooperative knowledge types. Receptive knowledge can be something like an example or an explanation, interactive knowledge can be something like a simulation or a multiple choice question, and cooperative knowledge types include discussions, disputations, group work and so on [16]. The knowledge types are compiled from the pedagogical literature since Comenius and thus allow to express most teaching and learning concepts that are used in western culture, like upfront teaching, programmed instruction, digital problem based learning or inquiry learning.

All knowledge objects with the same topic shape one concept container. Concept containers are the second line of the Web Didactic classification. The concept containers are described by the topic and their relations to other concept containers. These relations are typed relations. Mainly hierarchical and associative relations are distinguished. The topics of the concept containers and the typed links build a thesaurus.

Concept containers are combined into knowledge domains, which are the third line of the Web Didactics. For simplicity, a knowledge domain can be associated with a course. The course is described by a topic again. These topics can be related by typed links. Courses with the same topic can be understood as modules. Modules are on the forth line linked into curricula. Thus a network of domain specific thesauri is related within the classification. This way, the approach can be used for the individualization of modules and curricula, but this idea is behind the scope of this paper.

The classification of knowledge objects, concept containers and knowledge domains is combined with a classification of learning pathways. In other terms: we are using a poly-hierarchical classification system with domain specific thesauri. To do so, the concept containers and the knowledge objects can be arranged into multiple recommended learning pathways by the teacher. This is done by typed relations. These learning pathways are restricted to directed acyclic graphs. The learner can follow one of the recommended learning pathway or create his own learning style. If a learner explicitly of implicitly creates his own learning style, this can be applied to upcoming content based on the metadata vocabulary.

Recommendations for concept containers and knowledge objects are dynamically calculated while the learning process takes place. To do so, data from a learner model and learners log data are combined with the metadata the teacher created. The advantage of this ontology can be illustrated by the fact that all metadata are optional. Even if no metadata are given, recommendations and feedback can be calculated by using the learner log data. Still, with more metadata, the recommendation and feedback are more precise and cover more different situations. With teacher generated learning pathways, for example, learners can use these pathways. If the "blind"-field is filled, the recommendations can consider if the learner is blind or not, and so on.

In comparison with other approaches, the metadata set is small and simple. It thus can easily be applied in practice and used as required in the given context. Still, the metadata support individualized learning. Thus the freedom of the learner in the learning process is increased. This can be illustrated by the lost in hypertext phenomenon. If the learner uses a risky navigation style and looses orientation, the calculated recommendations and feedback can help him to continue his learning process.

At the same time, collaborative knowledge production is supported. It is, for example, possible to split the production of content by using the media types. One team might produce videos, while another team prepares readings and a third team creates forms for tasks. The same can be done on the level of concept containers. If, for example, a list of concept containers is covered by different people or teams, a common list of media and knowledge types that have to be covered in every concept container can be defined. In this case, the classification is used to coordinate the collaborative knowledge production.

2.2 A Pedagogical Ontology for Web Didactics

The pedagogical ontology[3] we developed is based on the Web Didactics vocabulary and classification. The development of the ontology is based on the following observations and experiences:

- Teaching and learning depends on heuristics that are based on authors and learners experiences and cultural backgrounds [17]
- The production of e-Learning material is costly.
- Managing large numbers of learning pathways is difficult for authors and learners.
- The production of learning environments is a professional activity that cannot be conceptualized into a rigid system [18].
- A reasonable granularity of learning material is required to be able to generate a comprehensible classification that is used for adaptations, recommendations and individual learning [19].

For the transformation of the Web Didactic concept into the pedagogical ontology, the context of the project is relevant. The INTUITEL system is intended as a plug in for existing learning management systems. All systems that are considered in the project (Moodle, Illias, Clix, Crayons) use a course as the highest aggregation level. Thus we used level 1 to 3 (knowledge objects, concept containers, knowledge domain) in the ontology only. While the learning management systems used in the project offer a suitable granularity, they are

---

[3] The pedagogical ontology developed within the INTUITEL project is available at http://www.intuitel.de/public/intui_PO.owl.

hardly flexible in terms of offered tools like forums, exercises and so on. Thus the ontology needs a flexible design that supports the adaptation of the vocabulary to the learning management used in a given situation. For practical usage it was also important that the ontology can easily be applied to existing courses.

When turning the metadata set and the vocabulary of the Web Didactic concept into an ontology, the heuristic characteristic needs to be considered. Due to the mentioned theoretical and practical reasons the ontology can not be created as a completed ontology, but needs to be created as an open ontology. To do so, we created the ontology and the INTUITEL system with the vocabulary from the Web Didactic [20,21], but designed the ontology and the software architecture in a way that keeps the possibility to change the vocabulary. Entries for media types, knowledge types or relation types can be added as needed and taken into account in the learning process. Our 15 years experience showed, that this is not happening very often. The media types vocabulary for example did not change in the last 15 years, since no new media types have been developed.

For the pedagogical ontology, we defined knowledge objects, typed links that form micro level learning pathways between the knowledge objects in the form of directed acyclic graphs, where nodes represent knowledge objects and edges represent specific types of micro learning pathways. We further defined concept containers and typed links between them in the form of directed acyclic graphs that form macro level learning pathways where edges represent specific types macro learning pathways, and knowledge domains as the basic entities. The ontology is the starting point for the Semantic Learning Object Model developed in INTUITEL.

2.3   The Semantic Learning Object Model (SLOM)

The Semantic Learning Object Model (SLOM) is a new metadata model developed in the INTUITEL project[4] to combine pedagogical and domain-specific knowledge with concrete learning material. SLOM complements existing and well-known eLearning formats such as Sharable Content Object Reference Model (SCORM)[5] and IMS-Learning Design[6] with semantic information that allows for a more intelligent and personalized (i.e., adaptive) processing of learning material in INTUITEL-enabled Learning Management Systems (LMSs). It serves as facilitating data infrastructure for the utilization and integration of externally hosted data in INTUITEL-compliant learning material.

The Semantic Learning Object Model (SLOM) is the format in which the INTUITEL system stores general information about courses, which is necessary to provide learning recommendations and feedback to learners. The SLOM format is implemented as a direct extension of the Pedagogical Ontology and defines how course information needs to be described in order to be compatible with the INTUITEL system. SLOM as a metadata format contains two ontologies for

---

[4] http://www.intuitel.de/.

[5] http://scorm.com/.

[6] http://www.imsglobal.org/learningdesign/.

a given course, the *Cognitive Map (CM)* and *Cognitive Content Map (CCM)*. The former is the description of topics in a domain of knowledge, while the latter describes the actual learning material of that course. A CM should be universally valid, meaning that CMs can be reused across different courses pertaining to a given topic. CCMs are, in contrast, specific to a given course since they enhance the actual learning content. SLOM as a storage format additionally contains the learning material on which the CCM is based in its original format. The SLOM specification prescribes the structure in which the given material has to be stored in order to be compliant. This entails three main pillars of information that should be compiled into the CM/CCM from the original content format:

1. *Topology*: Information about which elements are in the learning material as well as their topical coherences. In terms of INTUITEL, this means that the SLOM contains definitions for *Concept Containers (CCs)* and *Knowledge Objects (KOs)* of a given domain of knowledge.

   A Knowledge Object in INTUITEL is the smallest addressable part in an eLearning course, which is intended to provide insights into one distinct piece of knowledge. It is the anchor point for extending the content with metadata (e.g. knowledge type, expected learning time, etc.). Generally, this should represent about one screen page and correspond to roughly 5–10 min of learning time[7]. A KO always has a URI in context of the CCM it is embedded, which makes it possible to directly index the metadata it contains. Furthermore, if used in a running course, a LMS-ID uniquely identifies the element in the eLearning platform and, in context of a SLOM package, a SLOM-reference that links to the page in the package structure.

   Concept Containers on the other hand are structural components that allow for combining Knowledge Objects in different topics. It is possible to attach one KO to different CCs and thus create complex knowledge coherences across a course.

2. *Sequences*: Learning Pathways (LPs) on different levels that allow interlinking Knowledge Objects and Concept Containers. This is one of the main elements in the INTUITEL system as a whole and gives teachers the opportunity to compile their courses in different ways. On the topical level, *macro Learning Pathways (MLPs)* describe the sequence in which a learner should work through Concept Containers. On the content level, *micro Learning Pathways* (μLPs) describe how learners should work through Knowledge Objects. The latter has only to be done in a smaller context, meaning only inside a given complex of meaning. The total set of Learning Pathways results implicitly by combining MLPs and μLPs. So, although teachers only describe a relatively small number of pathways, the actual number of possibilities of working through a course is the product of them. Generally, learning pathways can be seen as a (set of) directed acyclic graphs.

---

[7] Naturally, this varies from element to element, but can be taken as a guide value for the creation of courses. Especially in context of different content types (e.g. tests, assignments, definitions, etc.) and media types (e.g. video, text, etc.), this is actually expected to vary.

3. *Background*: In addition to interconnecting elements, the PO and consequently also the SLOM format allows to describe these elements. This concerns the Knowledge Objects (KOs), which, in contrast to Concept Containers (CCs), have a real representation as a course. The respective content elements have properties as seen from a technological and didactical perspective. The former, for instance, concerns their size (e.g. 1 MB) or recommended screen resolutions (e.g. 1024×768 pixels), while the latter regards the educational purpose and background of the elements. This could, for example, be the difficulty level (e.g. beginner) or the type of knowledge it contains (e.g. different types of assignments).

A combination of these three pieces of information with learner data allows the INTUITEL system to create recommendations for appropriate learning objects and to produce feedback messages in that process. The more information that can be provided on the course, the more information can be integrated in this process.

2.4   Overview of Semantic MediaWiki

In this section, we provide a concise but non-exhaustive overview of the main language elements of Semantic MediaWiki systems[8]. This overview serves as basis for the subsequently following elaborations on the import and mapping declarations that need to be defined between the knowledge representation formalisms underlying Semantic MediaWiki and external ontologies such as the PO and SLOM as OWL ontologies (see also Sect. 4.2).

Semantic MediaWiki[9] is a free and open-source extension to the MediaWiki software[10] that allows for adding machine-readable semantic information in the form of semantic annotations to wiki articles. Semantic annotations are materialized in the form of *Categories, semantic Properties, Subobjects,* and *Concepts* and allow for complementing existing wiki pages with facts and explicitly defined relationships to related articles in a structured and meaningful way. Information represented as semantic annotations can be queried and aggregated in more sophisticated ways compared to articles that use the default elements defined by the MediaWiki language model.

Semantic MediaWiki was developed as a full-fledged framework to complement MediaWiki with functions found in knowledge management systems [23]. One of the main distinguishing features of Semantic MediaWiki compared to MediaWiki is the interoperability of the data created with it, as its underlying description framework is based on concepts, languages, and technologies defined by W3C semantic Web standards[11], the vision of which are to evolve the

---

[8] For a detailed introduction to Semantic MediaWiki and the unique benefits it adds to MediaWiki, we refer the reader to the official Semantic MediaWiki manual [22] or the related literature (e.g. [23,24].

[9] https://semantic-mediawiki.org/.

[10] https://www.mediawiki.org/wiki/MediaWiki.

[11] http://www.w3.org/2013/data/.

Web into a global data space of linked data sources, where RDF and common ontologies serve as interoperability infrastructure (cf. [25]). This interoperability infrastructure allows external applications to use and integrate data created with a Semantic MediaWiki in a controlled and meaningful way. It also enables the integration of semantic search capabilities in Semantic MediaWiki systems.

Semantic MediaWiki (SMW) in general does not define a new canonical data or description format since the logical model that builds the basis of its knowledge representation formalism is to a large extend based on the Web Ontology Language (OWL). This reliance enables a direct mapping (cf. [24]) of baseline Semantic MediaWiki elements to OWL language elements (see also Table 1):

- *Categories* in a Semantic MediaWiki system are represented as named classes in OWL ontologies; ontology classes, on the other hand, can be directly mapped to categories in Semantic MediaWiki.
- *Articles* created within Semantic MediaWiki are treated as individuals of an ontology and hence as members of the classes that represent the Semantic MediaWiki categories a page is related to.
- *Properties* are the Semantic MediaWiki pendant to roles in Description Logics and properties in OWL.

Table 1. Direct mapping of OWL language elements to Semantic MediaWiki

OWL Language Element	Semantic MediaWiki
OWL Individual	Normal article in the default namespace
`owl:class`	Article in the `Category` namespace
`owl:ObjectProperty`	Article in the `Attribute` namespace
`owl:DatatypeProperty`	Article in the `Attribute` namespace with [[has Type::...]] declaration
OWL class expression	Article in the `Concept` namespace[a]

[a]Such pages are exported as OWL class expression (see https:// semantic-mediawiki.org/wiki/Help:Concepts)

In contrast to OWL properties, i.e., roles in description logic, SMW does not distinguish between *object* and *datatype properties* respectively concrete and abstract roles. Both elements are mapped to articles in the `Attribute`-namespace where Semantic MediaWiki's RDF Exporter (see Sects. 4.4 and 5.2) determines a property's type in terms of OWL DL depending on the occurrence of a [[has type::...]] declaration. If such a declaration is found on a property's article page, then the property is treated as an `owl:DatatypeProperty` in the export and its value is mapped to the value space of the respective datatype. An external reasoner can then check whether the given value corresponds to the range definition of its associated OWL datatype property.

Unlike OWL, which is built on the *non-unique name assumption* (cf. [26,27]) and allows identical entities to be referred to via different IRIs, SMW interprets articles with different IRIs as different individuals per default. However, in order to state that two articles with different IRIs are identical, SMW adopts the concept of *redirects* from MediaWiki to express equivalence between differently named categories, properties, and articles. In terms of OWL, the concept of redirects resemble equivalence assertions between individuals using `owl:sameAs` as well as among classes and properties expressed through `owl:equivalentClass` and `owl:equivalentProperty`. Table 2 summarizes the different types of equivalent expressions in OWL and Semantic MediaWiki:

Table 2. Expressing equivalence in OWL and Semantic MediaWiki

OWL Language Element	Semantic MediaWiki
`owl:sameAs`	`#REDIRECT [[{pagetitle}]]` —on normal article pages
`owl:equivalentClass`	`#REDIRECT [[{pagetitle}]]` —articles in the **Category**-namespace
`owl:equivalentProperty`	`#REDIRECT [[{pagetitle}]]` —articles in the **Attribute**-namespace

SMW also allows for the declaration of value spaces to restrict a property's value range to a list of allowed values the property may hold. This restriction might be complemented by additional normative and non-normative information. However, normative information can only be specified in an informal way and hence prevents consistency checking by a formal reasoner (which is possible, for instance, in OWL ontologies and common OWL/DL reasoners).

3   Related Work

The fields of ontology engineering, semantic Web technologies and Linked Data are being strongly connected in order to provide intelligent applications that can support learners in organizing their studies and connecting adequate learning resources in pedagogically meaningful learning paths. Many authors have therefore stressed the importance of Linked Data and semantic technologies on e-learning as well as the tools for transforming existing, legacy data into Linked Data [5,7,28,29]. This resulted in developments of so-called *Semantic Learning Management Systems (SLMS)* and *Web Science Semantic Wikis (WSSW)* to exceed the self-contained perspective of current semantic MediaWiki systems in terms of openness for external semantic queries [5]. Such a feature allows content to be collaboratively authored and exposed as Linked Data in an ad-hoc manner and become incorporated into other semantic data structures on-the-fly. This not only requires semantic Web languages such as RDF/S and OWL as interoperability infrastructure but also the authoring of pedagogically meaningful content annotations. Li et al. [7], for instance, demonstrated how learners and content authors can benefit from a collaborative elearning environment backed by Semantic MediaWiki in terms of authoring, access, sharing, and reuse.

The importance of content authoring for the acceptance of educational systems is analyzed by several works (e.g. [3,6,10,30]). Sosnovsky et al. [3] present a topic-based knowledge modeling approach, which was inspired by instructional design practice and claims that *"domain model does not have to be very detailed to ensure the effective adaptive behavior and usability of the system"*. While reusability is ensured, the aspect of collaborative authoring is not considered.

The same can be found in [30]. The authors introduce an ontology-aware authoring system for learning designs. It is designed in compliance with some international standards (SCORM and LOM) in order to enhance shareability and reusability of learning designs among users, but nothing is stated about collaborative authoring. Besides, the system collects and searches learning resources suitable to the authors. However the tool does not have the functions to edit metadata.

Holohan et al. [10] present a set of software tools aimed at supporting authoring, management, and delivery of learning content that build on semantic Web technologies for knowledge representation and content processing. A key feature of the system is the semi-automatic generation of standard e-learning and other courseware elements through graph transformations on underlying ontologies. Their system also offers features such as adaptivity in terms of students' learning track guidance, ontology engineering, as well as dynamic content delivery based on configurable navigation pathways. Information regarding an ontology-based representation of learning pathway semantics or the pedagogical concepts to which their approach pertains is not provided. The aspect of collaboration is also not addressed in their work.

The potential of semantic Web based knowledge representation frameworks for the development of learning content along seven different application domains for ontologies is surveyed in [6]. Their research also corroborates the importance of ontologies for content adaptation, content assembly, and content reuse.

Other works (e.g. [2,4]) specifically emphasize the multidisciplinary character of the content creation process and underline the relevance of supporting collaboration due to the different roles and tasks involved. The integrated framework developed by Dodero et al. [2] supports the collaborative authoring and annotation of learning objects and has been realized in form of an Eclipse RCP application. However, the collaboration module offers functionalities for negotiating and evaluating annotation proposals although in a style different from that found in today's Wiki systems. Extending current semantic MediaWiki systems with additional collaborative editing features such as offline work support and multi-synchronous edits is proposed by Rahhal [11]. Although adaptivity is not addressed, their work extends the presented approach in useful ways.

Brut et al. [4] motivate the usage of semantic Web technologies for addressing the challenges of accessing learning content not only across e-learning platforms but also across Web applications, which resulted from the intentional shift of current e-learning solutions towards decentralization and inter-institutional collaboration. The proposed method combines semantic technologies with TF-IDF-indexing, Latent Semantic Indexing, and WordNet-based processing for extending the IEEE LOM standard [31] with ontology-based semantic annotations. While their approach remains ontology-independent, the aspects of collaborative authoring of annotations is not particularly addressed.

Development methodologies to encourage and support domain experts in developing ontologies for the annotation of learning content were proposed by [8,32]. Unlike the simple is-a relationship provided by many ontologies in the educational context [8], ontologies that provide a richer and more expressive set of relationship types are required. The authors also revealed that a separation of encoded knowledge into concept space and educational content space supports utilization flexibility. These aspects are satisfied by SMW's knowledge representation formalism (and those of all DL ontologies) as encoded knowledge is separated into assertional and terminological knowledge and exported content can be mapped to more expressive ontologies to utilize the full feature set of enhanced LMSs—as demonstrated by the present work.

4   Approach

After having described the expressive means of Semantic MediaWiki's knowledge representation formalism in Sect. 2.4, we now specify their semantics in terms of the OWL DL part of the Web Ontology Language. We show how pedagogically relevant terms can be mapped to Semantic MediaWiki's language elements and vice versa so that course content can be exported in a format that is compliant to the Pedagogical Ontology and Semantic Learning Object Model. We first discuss the limitations of Semantic MediaWiki's knowledge representation formalism compared to the rather expressive OWL DL language upon which the PO is defined (see also [24]). We then describe the process of importing PO and SLOM elements into Semantic MediaWiki using import and mapping declarations and demonstrate how course designers can collaboratively create semantically annotated learning material. In the last part, we expound how such content can be exported in a format that is compliant to the Pedagogical Ontology and SLOM in order to make them available as Linked Data.

4.1   Overview

The main objective of the presented approach is to enable different roles in the content creation process to use pedagogically expressive annotations in a Semantic MediaWiki system. Terms from external ontologies are extracted and imported into a Semantic MediaWiki system using *import declarations*. Imported terms are declared on special import pages (see Sect. 4.3). Imported terms can then be mapped to Semantic MediaWiki-specific terms (the individual terms of content authors) using import declarations (see Sect. 4.4) and serve as *associated terms*. Once import and mapping declarations are defined, these terms can then be used for creating and annotating learning material (see Sect. 4.6). Once annotated learning material is exported through the RDF Export facility, SMW-specific terms are automatically resolved and remapped to their associated ontology terms (so that the original ontology terms are contained in the exported data). Learning content and its annotations are exported as Resource Description Framework (RDF) data. However, in order to deploy exported learning content in a Learning-Management-System (LMS), the URIs of contained media files need to be dereferenced in order to retrieve the concrete media content and incorporate it in a Content Package (see Sects. 2.3 and 7) that can be processed by an INTUITEL-enabled LMS (Fig. 1).

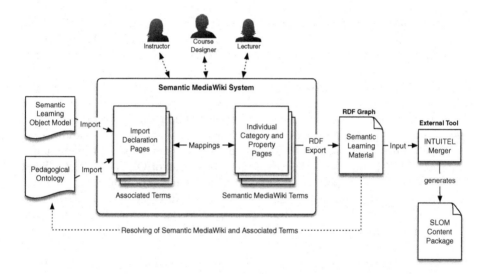

Fig. 1. Overview of the presented approach including subsequent steps

4.2 Limitations of Semantic MediaWiki's Knowledge Representation Formalism

This section answers two questions: First, it elucidates why the PO and SLOM can not be directly converted into Semantic MediaWiki although both knowledge representation formalisms are defined on the basis of Description Logics. Second, it discusses reasons why a universal mapping between learning material created and annotated with Semantic MediaWiki and learning material annotated on the basis of the Web ontologies such as the PO and SLOM can not exist.

The axioms that constitute an OWL ontology—irrespectively of the profile— can be separated into axioms representing *terminological*, i.e., schema knowledge and axioms that encodes information about individuals, i.e., *assertional knowledge* (see [33–35]). Although this distinction has no model-theoretic meaning, it is useful for modeling purposes [35]. On TBox and RBox level, ontologies axiomatically define a definite set of terms and constraints, the interpretation of which is determined by the formal semantics of the underlying ontology language in which an ontology is represented (see [36,37]). The accurateness by which such constraints can be expressed depends on the logical theory and hence on the Description Logic upon which an ontology language is defined. The logical theory also determines the conclusions (logical entailments) that can be deduced from a formal interpretation of the elements semantics.

The PO and SLOM are encoded using the ontology language OWL DL, which is based on the $\mathcal{SHOIN(D)}$ Description Logic that exhibits NExpTime-reasoning complexity while still being decidable [26,38,39]. OWL 2 EL [39], which is defined on the family of $\mathcal{EL}$-Description Logics and comparable to Semantic MediaWikis knowledge representation formalism, employs PTime-complete complexity and allows for polynomial time reasoning[12]. More expressive languages such as OWL Full contain a richer set of language elements for defining logical axioms, however, at the cost of being undecidable.

Semantic MediaWiki adopts the set-based semantics of OWL for classes and roles (see [40]) and exhibits features such as equality reasoning and reasoning on the transitivity closures on category and property hierarchies. However, most of the expressiveness incorporated in OWL DL is not available in SMWs knowledge representation formalism. For scalability and consistency reasons, the language model of SMW is built on a less expressive fragment of OWL DL that allows for polynomial time reasoning on large corpuses of instance data at the cost of excluding some of the formal semantics well-known in OWL DL (cf. [24]).

Therefore, not all elements and ontological (TBox) constraints defined in the PO can be directly expressed in form of SMW elements. For instance, the SMW language model does not define elements for explicitly expressing *inverse properties* or *disjointness*. In the latter case, this means that in SMW, it is not possible to define a disjointness restriction on category level that expresses that two categories do not hold any page in common, i.e., an article cannot belong to both categories at the same time. For instance, the formal semantics defined in the PO TBox that the classes `ConceptContainer`, `KnowledgeDomain`, and `KnowledgeObject` are mutually disjoint cannot be expressed using Semantic MediaWikis knowledge representation formalism.

However, SMW provides some basic OWL DL language features that enable the formulation of *complex class expressions*, i.e., defining class membership constraints by means of logical axioms such that different requirements must hold for an individual to become member of a class. The language feature *Concept* enables the declaration of *dynamic categories* which contains only those articles that hold specific properties to pages belonging to another category or properties with specific values. Unlike OWL DL, concepts can not be used in combination with quantifiers or cardinality constraints. SMW does also not define means for expressing restrictions on the formal semantics of the data being annotated as it would be possible with OWL DL.

---

[12] cf. http://www.w3.org/TR/owl2-profiles/#Computational_Properties.

In consequence, SMW's knowledge representation formalism does not allow for formally evaluating the *logical consistency* of course material being annotated with terms imported from the PO and SLOM, as it would be the case with full-fledged ontology editors such as Protégé[13] and standard DL reasoners. That means that any inconsistency that might be introduced by a course designer cannot be automatically detected by a Semantic MediaWiki at design time of a course, but have to be dealt with at later stages, e.g., by external components since the limited expressiveness of Semantic MediaWikis knowledge representation formalism restraints users from the peril to unintentionally introduce formal inconsistencies on ABox, RBox, and TBox level to their ontologies [24].

In addition to the differing expressivity aspect of the underlying knowledge representation formalisms, a second aspect that need to be taken into consideration for exposing collaboratively created learning material as Linked Data is *schema compliance*. In contrast to ontological domain specifications, SMW does not make any assumptions regarding pre-existing classification schemes or vocabularies used for the description of domain knowledge. This means that elearning content creators have the freedom to individually define the vocabulary terms depending to the universe of discourse to which their learning material pertains. Due to the difference in terms of schema compliance between the PO and SMW[14] a universally valid approach can not be realized. Instead, import and mapping declarations for existing vocabularies or classification schemes need to be defined on an individual basis. However, such terms can be reused and refined by different content authors. This approach fits well into the given scenario as Semantic MediaWiki in general is an appropriate tool for authoring the instance data of complex ontologies since these are subject to more frequent changes compared to the rather stable terminological knowledge of ontologies.

4.3   Creating Vocabulary Import and Mapping Declarations

In a first step, the set of external vocabulary terms that should be available for content authors in a Semantic MediaWiki system must be declared using the special page `MediaWiki:SMW_import_{namespace}`. The special page `smw_import_{namespace}` contains a list of those vocabulary terms that should be imported and for which mapping declarations could be made; the elements can be individually chosen. The page belongs to the Mediawiki namespace and has the prefix `smw_import_`. It can only be created by users with administrator status and involves the declaration of an individual namespace in form of a `qname` (indicated in the `{namespace}`-part) to uniquely identify and reference imported terms in the mapping declarations of the associated Semantic MediaWiki terms.

---

[13] http://protege.stanford.edu/.

[14] The PO allows for the description of domain knowledge in a way so that an INTUTEL-enabled system is able to process such data and provide sophisticated services in the form of individual recommendations; Semantic MediaWiki, in contrast, does not exhibit any predefined or default schema nor impose any restrictions on the definition of individual schema information—apart from those imposed by the underlying data model.

```
1    http://www.intuitel.eu/public/intui_PO.owl#|
2    [http://www.intuitel.de/public/intui_PO.owl
3    Pedagogical Ontology of the INTUITEL Project]
4     AbstractOrientation|Category
5     ActionReceptive|Category
6     AddressSource|Category
7     AnimationPresentation|Category
8     [...]
9     containsConceptContainer|Page
10    containsKnowledgeObject|Page
11    containsLearningObject|Page
12    hasBottomUpLikeRelation|Page
13    hasCharacterizingObjectProperty|Page
14    hasChronologicalLikeRelation|Page
15    hasFromNewToOldLikeRelation|Page
16    hasFromOldToNewLikeRelation|Page
17    hasRecommendedAge|Type:Number
18    isLinkedWithSlomPackageElement|Type:Text
19    isLinkedWith|Type:Text
20    isSuitableForBlind|Type:Boolean
```

Fig. 2. Excerpt of import declarations for Pedagogical Ontology's Elements

Once the special import page is created, it can then be populated with import declarations. For making the PO and SLOM elements available in a SMW system, the import declarations of the import page must follow a specific notation and structure, which is illustrated in Fig. 2[15].

For each vocabulary that is to be imported into a Semantic MediaWiki instance, a base URI must be specified. In most Semantic Web vocabularies, it is common practice to specify terms as fragments of the vocabularys URI. For instance, the URI of the class KnowledgeDomain defined in the PO is

http://www.intuitel.eu/public/intui_PO.owl#KnowledgeDomain.

Before an agent can retrieve a machine-processable representation of the given concept, it first needs to strip-off the fragment part from the vocabularys URI and then de-reference the base URI (see [41]).

The following lines in the special import page declare each vocabulary element that will be imported and might be reused. This is the main part of the special import page and mandatory for declaring the mappings between the associated ontology terms and the individually defined SMW terms. Each line in the main part of the special import page starts with a whitespace followed by the specific name of the element (for most vocabularies, this is the fragment of the elements URI). The text after the pipe symbol ('|') declares the context in

---

[15] For readability reasons, we sorted the elements alphabetically and separated category and page import declarations; we also included both element types (PO and SLOM) is this excerpt although we advocate to separate PO and SLOM elements for maintenance reasons.

which an element can be used in the wiki. This part is important, as Semantic MediaWiki distinguishes between classes and properties to be imported: classes can only be used as categories; OWL object-, data-, and annotation properties can only be mapped to Semantic MediaWiki properties (see Table 1). The default type assignment for object properties is `Page`. For OWL datatype properties, a datatype must be explicitly stated using the `Type:{some datatype}` declaration, otherwise the default datatype `Page` is set. Vocabulary terms that should be imported as categories must be declared as `Category` using the category namespace identifier.

In order to separate the default PO elements from the SLOM elements contained in the PO, it is useful to create a separate import declaration page (e.g. `smw_import_slom`) to hold only those import declarations pertaining to the SLOM-specific datatype properties. Since all SLOM elements are defined as OWL datatype properties in the PO, each import declaration contains a type declaration that refers to a specific SMW datatype (see *List of Semantic MediaWiki Datatypes*[16]).

4.4   Creating Mapping Declarations for Associated Terms

The import vocabulary function of Semantic MediaWiki allows for the declaration of mappings between individually defined Semantic MediaWiki-specific terms and terms for external vocabularies, the so-called *associated terms*. In a second step, those mapping declarations need to be added to the Semantic MediaWiki property and category pages that are to be mapped to the imported vocabulary elements. This is done using the following statement on the individual property or category page:

```
[[imported from::{namespace}:{element_name}]]
```

The special property `imported_from` signals SMW that the element onto which page this declaration was added actually refers to the element specified by its namespace and name (e.g. `foaf:knows`) after the double colons ':::'. Basically, all elements from external vocabularies in general and the PO specifically can be imported as described above. For instance, a mapping declaration on a category page that refers to the `intuit:KnowledgeDomain` class of the PO can be added to the category page in the following way:

```
[[imported from::intui:KnowledgeDomain]]
```

By interpreting this statement, the system can relate the category page to the associated term since it has been made available previously on the import declaration page. More information about the import of external vocabularies into a Semantic MediaWiki system can be found on the help page of the import vocabulary function in the Semantic MediaWiki Manual[17].

---

[16] http://semantic-mediawiki.org/wiki/Help:Properties_and_types#Datatypes_for_ properties.

[17] https://semantic-mediawiki.org/wiki/Help:Import_vocabulary.

The import and mapping declarations also play a crucial role in the export of Semantic MediaWiki-authored course material as they built the basis for mapping the individual vocabulary terms defined by the course creator back to the respective elements of the PO and SLOM. This means, that the URIs of those categories and properties for which mappings have been declared are replaced by the URIs of the elements the mapping declarations refer to. This is an important aspect as terms being defined in a SMW system are local to it, i.e., when the URI of an element that is defined inside a Semantic MediaWiki system is exported to RDF via SMW's RDF export functionality, the element's namespace per default resembles the namespace of the MediaWiki system from which the element was exported. Technically, the associated terms from external vocabularies work like any other term in the Semantic MediaWiki, but the RDF data that are created when selected pages from the wiki are exported directly contain resolved PO and SLOM terms. In consequence, the Semantic MediaWiki terms are replaced during the export by its *associated terms* for which mappings have been declared.

4.5   Integrating Elements from External Ontologies

Although we have described the import and mapping declarations for the PO and SLOM elements exclusively, these elaborations can also be used for integrating arbitrary OWL ontologies in Semantic MediaWiki and making their elements available. For instance, the Pedagogical Ontology uses terms from the *Dublin Core Ontology*[18] for annotating its elements with title and description information using `dc:title` and `dc:description`. In order to maintain compliance with the annotation vocabulary of the PO, we recommend to also import these terms into a Semantic MediaWiki and use the corresponding wiki pages for the annotation of individually created learning material (CM and CCM). When the learning material is exported via the RDF Exporter (see Sect. 5.2), external tools can interpret and process these annotations along with the standard Dublin Core annotations contained in the PO, since their Semantic MediaWiki-specific namespaces will be replaced by the URIs of the associated terms (see previous section), even in case the terms are named differently.

4.6   Defining Individual Annotations

All elements from the imported external ontologies or vocabularies (PO, SLOM, Dublin Core, etc.) can be used for annotate content in the same way that any SMW property. Annotations in SMW are defined by specifying the value for the element after the colon ':', that is, [[{element_name} : ...]], where {element_name} is the SMW property that has been mapped to the imported vocabulary element.

First of all, each page in a Semantic MediaWiki system should be associated to a Semantic MediaWiki category, for example:

---

[18] http://dublincore.org/schemas/rdfs/.

```
[[Category:ConceptContainer]]
```

where `ConceptContainer` should be mapped to the corresponding term of the PO (whose vocabulary must be imported beforehand into a Semantic MediaWiki system as detailed in Sect. 4.4).

Next, the page could be annotated making use of external vocabulary elements. For example, the following declaration in the page KD_A:

```
[[containsConceptContainer::CC_1|Name of the CC_1]]
```

relates CC_1 with KD_A by the property `Property:containsConceptContainer` (imported from the PO). Besides, a link to the page is included in the actual wiki page (the text following the pipe symbol '|' is shown rather than the real name of the linked page).

The `#set` statement can also be used for specifying properties' values in order to not show them on an article page: `#set:{element_name}={element_value}`. The following excerpt illustrate its usage in adding SLOM-specific metadata to an article page

```
{{#set:isSuitableForBlind=false}}
```

where the property `isSuitableForBlind` was defined as a new Boolean-type SMW property and mapped to the corresponding SLOM element.

5   Use Case: Adding Pedagogical Semantics to an E-Learning Course at the University of Valladolid

This section provides a description of the steps needed to model eLearning Course Content using a collaborative tool like Semantic MediaWiki. The specific objective is annotating existing learning content in order to make it into INTUITEL-compliant learning material. A real course about "Network Design" held at the Telecommunications School of the University of Valladolid has been used as use case[19].

This course focuses on different design aspects of four types of networks. There are two main alternative approaches to learn this course: (1) studying the different types of networks with their design considerations separately; (2) organizing the content hierarchically, in which the design aspects are considered as main topics, analyzing each aspect per type of network. Therefore, the cognitive map defined by the teachers of this course includes two macro learning pathways, using two of the MLP types pre-defined in the pedagogical ontology. Figure 3 shows part of the cognitive map, which has in total 24 CCs.

The above mentioned first approach corresponds to the pathway labelled as `hasLogicalNextStep` while the second one is labelled as `hasHierarchicalNextStep`. For example, arrows of type `hasLogicalNexStep` guide from CC Wan

---

[19] Access to the Semantic MediaWiki system in which the course is modeled can be granted on request; please contact zander@fzi.de.

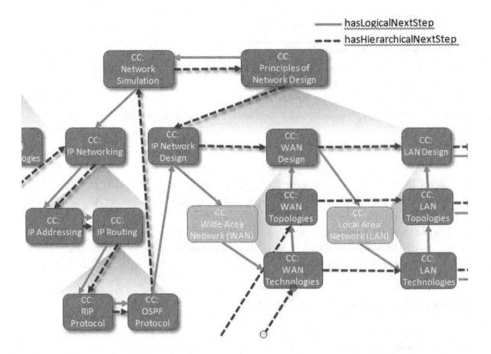

Fig. 3. Excerpt of the cognitive map of the course "Network Design"

Technologies to CC Wan topologies and, then, to CC Wan design, studying consecutively all the different aspects for the type of network WAN. However, in the pathway hasHierarchicalNextStep, the CC LAN technologies is located after CC WAN technologies, following the approach of consecutively studying the different aspects, technologies in this case, for each type of network (WAN, LAN, etc.).

The main page of the course represents the Knowledge Domain. This page contains an index or table of contents with 24 topics or lessons of the given course (the concept containers). With this approach, the CCs contained in a KD (e.g. NetworkDesign) can be specified in a straightforward way as illustrated in the SMW markup code excerpt displayed in Fig. 4:

```
1   [[Category:KnowledgeDomain]]
2
3   [[containsConceptContainer::ccPrinciplesOfNetworkDesign
4     |Principles of Network Design]]
5   [[containsConceptContainer::ccNetworkSimulation|Network Simulation]]
6   [[containsConceptContainer::ccIPNetworking|IP Networking]]
```

Fig. 4. Excerpt of the Semantic MediaWiki page kdNetworkDesign

Each Learning Object (KD, CC, KO) has its own page in a Semantic MediaWiki system. Pages can be created and annotated by different authors. When creating the page, it must be associated with the correct Semantic MediaWiki category. Therefore, three categories have been created together with their correspondent mappings to LO elements for the three different types of learning objects (a) `Category:KnowledgeDomain`, (b) `Category:ConceptContainer`, and (c) `Category:KnowledgeObject`.

Once the learning material is semantically annotated, it can be exported to OWL/RDF format by means of Semantic MediaWiki's RDF Export facilities.

5.1 Annotating E-Learning Course Content

This section shows some examples of how imported elements can be used for annotating real course content.

The first example corresponds to a wiki page for a CC, which is equivalent to a unit of the course. The CC page contains the metadata annotations as well as a list of all the course KOs. Figure 5 depicts how the imported elements can be used for annotating a wiki page that corresponds to a CC:

```
1   [[Category:ConceptContainer]]
2   =[[title::Concept Container -- Network Simulation]]=
3
4   <!-- Description of the CC's content -->
5   [[description:: This Concept Container ...]]
6
7   Contents of this topic:
8   *[[containsKnowledgeObject::KO_PresentationSimulationOPNETModeler |
9     Presentation about Simulation with OPNET Modeler]]
10  *[[containsKnowledgeObject::KO_LaboratoryExerciseOPNETModeler |
11    Laboratory exercise with OPNET Modeler]]
```

Fig. 5. Excerpt of the Semantic MediaWiki page `ccNetworkSimulation`

First of all, the page is associated with the category `ConceptContainer`. Next, the LO is annotated making use of the DC elements `title` and `description`. All these elements used for annotation should have been previously imported, as described in previous sections. Finally, the content, i.e., the knowledge objects the CC consists of are specified in form of a list of KOs. Each KO is linked and associated by the property `containsKnowledgeObject`, which previously should be mapped to its associated PO term (see Sect. 4.4).

A page of type concept container should also contains the CCs it links to, while using the adequate properties in order to form the correct INTUITEL *macro learning pathways*. In the current CM—as mentioned before—the following two Macro Learning Pathways have been defined:

- `hasLogicalNextstep` as a sub property of `hasFromOldToNewLikeRelation`
- `hasHierarchicalNextstep` as a sub property of `hasTopDownLikeRelation`

where `hasFromOldToNewLikeRelation` and `hasTopDownLikeRelation` are defined in the INTUITEL PO. Since SMW supports the definition of semantic sub properties, these two custom macroLP properties can be directly mapped to SMW-compliant properties (that might share identical labels). The range of those SMW-compliant macroLP properties are SMW pages that belong to the category CC. Therefore, two new SMW properties (and the correspondent SMW articles) have been created:

- `Property:HasLogicalNextstep` and
- `Property:hasHierarchicalNextstep`

Then, they must be defined as sub properties of two new SMW properties, which map to the imported PO properties. For example, the page in the property namespace `Property:HasLogicalNextstep` contains the following statements (corresponding to two special properties of the wiki):

```
[[Has type::Page]]
[[subproperty of::HasFromOldToNewLikeRelation]]
```

where `HasFromOldToNewLikeRelation` is another specific SMW property mapped to the imported PO term. The same should be done for each custom Macro Learning Path property. Then, they can be used in pages of CCs. For example, the page of declarations `ccNetworkSimulation` includes:

```
[[hasLogicalNextstep::ccIPNetworking]]
[[hasHierarchicalNextstep::ccPrinciplesOfNetworkDesign]]
```

The second example corresponds to a wiki page for a KO. A KO page will contain the CMM metadata annotations as well as the content of the KO or references to resources as PDF files. For example, the excerpt shown in Fig. 6 is part of the declarations page of `KO_LaboratoryExerciseOPNETModeler`. All the properties must be created the first time by defining SMW properties and then mapping them to the elements of the imported vocabulary.

A KO page should also contain links to other KOs, but using the adequate properties in order to form the INTUITEL *micro learning pathways*. There are no custom micro-level relations in the CMM, but properties defined in the INTU-ITEL PO. Then, once the SMW properties have been created and mapped to the correspondent imported PO properties, they can be used in pages of KOs in order to relate KOs to form a micro learning path.

5.2   Exporting the Semantically Annotated E-Learning Content

The INTUITEL-compliant learning material can be exported into OWL/RDF format by means of SMW facilities. The Semantic MediaWiki's RDF Export function is called by using the special RDF Export page. It generates an

```
1    [[Category:KnowledgeObject]]
2
3    {{#set:hasKnowledgeType=ktStepByStepGoodPractice}}
4    {{#set:hasMediaType=mtVideoPresentation}}
5    {{#set:hasEqfLevel=6
6     |hasEstimatedLearningTime=12 minutes
7     |hasLanguage=ES
8     |isSuitableForBlind=false
9     |isSuitableForDeaf=false
10    |isSuitableForMute=true
11    |hasRecommendedAge=10}}
```

Fig. 6. Excerpt of the article page KO_LaboratoryExerciseOPNETModeler

OWL/RDF document with the import and mapping declarations for the articles pages of the individual elements. The export function also assigns URIs to all articles that are exported, and replaces the URIs of those Semantic MediaWiki TBox elements for which a mapping declaration to their corresponding, i.e., associated elements from the PO has been defined.

The RDF Export function generates an OWL/RDF document with regard to the import and mapping declarations for the article pages of the individual elements. The export function also assigns URIs to all articles that are exported, and replaces the URIs of those Semantic MediaWiki TBox elements for which a mapping declaration to an associated element from the PO has been defined. The pages corresponding to the examples of the previous section have been exported using the special RDF Export page. Next we are going to show how some annotations are represented in the exported files.

Figure 7 displays an excerpt of the exported RDF file for the wiki page ccNetworkSimulation represented as Manchester OWL Syntax[20] (see the origin annotated wiki page in Fig. 5). The lines

```
1    Individual: wiki:CcNetworkSimulation
2      Types:
3        intui:ConceptContainer,
4        swivt:Subject
```

allows to identify this object as an instance of the class ConceptContainer defined in the published INTUITEL PO. This is the exporting result of the annotation in line 1 of the code excerpt of Fig. 5.

With reference to the wiki article about the CC NetworkSimulation, the exported RDF document does not miss any relevant information.

For example, the 2 KOs annotated in the origin wiki page (lines 8 and 10 of the code excerpt in Fig. 5) can be retrieved from this RDF file as the property intui:containsKnowledgeObject links the CC with the 2 wiki

---
[20] http://www.w3.org/TR/owl2-manchester-syntax/.

```
1   Prefix: dc: <http://purl.org/dc/elements/1.1/>
2   Prefix: owl: <http://www.w3.org/2002/07/owl#>
3   Prefix: rdf: <http://www.w3.org/1999/02/22-rdf-syntax-ns#>
4   Prefix: xsd: <http://www.w3.org/2001/XMLSchema#>
5   Prefix: rdfs: <http://www.w3.org/2000/01/rdf-schema#>
6   Prefix: wiki: <http://kalmar30.fzi.de/index.php/Spezial:URI-Aufl%C3%B6ser/>
7   Prefix: intui: <http://www.intuitel.eu/public/intui_PO.owl#>
8   Prefix: swivt: <http://semantic-mediawiki.org/swivt/1.0#>
9
10  Ontology: <http://kalmar30.fzi.de/index.php/Spezial:RDF_exportieren/CcNetworkSimulation>
11
12  Import: <http://semantic-mediawiki.org/swivt/1.0>
13
14  Annotations: swivt:creationDate "2015-09-08T11:51:24+02:00"^^xsd:dateTime
15
16  AnnotationProperty: swivt:creationDate
17  AnnotationProperty: rdfs:isDefinedBy
18  AnnotationProperty: rdfs:label
19
20  Datatype: rdf:PlainLiteral
21  Datatype: xsd:string
22  Datatype: xsd:dateTime
23  Datatype: xsd:double
24  Datatype: xsd:integer
25
26  ObjectProperty: wiki:HasLogicalNextStep1
27  ObjectProperty: wiki:HasHierarchicalNextStep1
28  ObjectProperty: swivt:page
29  ObjectProperty: intui:ContainsKnowledgeObject
30
31  DataProperty: swivt:wikiPageModificationDate
32  DataProperty: swivt:wikiNamespace
33  DataProperty: dc:title
34  DataProperty: dc:description
35  DataProperty: swivt:wikiPageSortKey
36  DataProperty: swivt:creationDate
37  DataProperty: wiki:Modification_date-23aux
38
39  Class: intui:ConceptContainer
40  Class: swivt:Subject
41
42  Individual: wiki:CcNetworkSimulation
43
44    Annotations:
45      rdfs:label "CcNetworkSimulation",
46      rdfs:isDefinedBy
47          <http://kalmar30.fzi.de/index.php/Spezial:RDF_exportieren/CcNetworkSimulation>
48
49    Types:
50      intui:ConceptContainer,
51      swivt:Subject
52
53    Facts:
54      wiki:HasLogicalNextStep1  wiki:CcIPNetworking,
55      intui:ContainsKnowledgeObject  wiki:KO_ExampleOPNETModeler,
56      intui:ContainsKnowledgeObject  wiki:KO_LaboratoryExerciseOPNETModeler,
57      swivt:page <http://kalmar30.fzi.de/index.php/CcNetworkSimulation>,
58      intui:ContainsKnowledgeObject  wiki:KO_VideotutorialOPNETModeler,
59      intui:ContainsKnowledgeObject  wiki:KO_PresentationSimulationOPNETModeler,
60      wiki:HasHierarchicalNextStep1  wiki:CcPrinciplesOfNetworkDesign,
61      dc:title  "Concept Container - Network Simulation"^^xsd:string,
62      swivt:wikiNamespace  0,
63      swivt:wikiPageSortKey  "CcNetworkSimulation"^^xsd:string,
64      dc:description  "This Concept Container contains information about Network Simulation
65          as a tool for designing all type of networks."^^xsd:string,
66      swivt:wikiPageModificationDate  "2014-03-05T09:36:46Z"^^xsd:dateTime
67
68  Individual: wiki:KO_VideotutorialOPNETModeler
69  Individual: wiki:CcPrinciplesOfNetworkDesign
70  Individual: wiki:KO_PresentationSimulationOPNETModeler
71  Individual: wiki:KO_ExampleOPNETModeler
72  Individual: wiki:KO_LaboratoryExerciseOPNETModeler
73  Individual: wiki:CcIPNetworking
```

Fig. 7. Excerpt of the exported Concept Container ccNetworkSimulation

pages containing the relevant data of those KOs (lines 56 and 59 in Fig. 7). Besides, the specific wiki properties `property:HasHierarchicalNextstep` and `property:HasLogicalNextstep` are used in order to locate this CC in the different Macro Leaning Pathways it belongs to:

```
1   Individual: wiki:CcNetworkSimulation
2      [...]
3      Facts:
4         wiki:HasLogicalNextStep wiki:CcIPNetworking,
```

Therefore, this shows how the approach could be used for collaborative definition of different learning paths. The same wiki content can be studied in different sequences, defined by the macro learning paths. Students, by themselves or guided by an intelligent tutor systems like INTUITEL, can select different learning paths thanks to the previously wiki annotations.

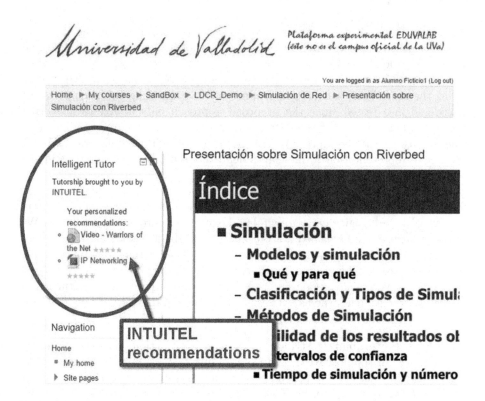

Fig. 8. Recommendations computed by the INTUITEL system based on the exported learning material for the course Network Design

Figure 8 shows an example of the recommendations provided by the INTUITEL system when using this INTUITEL-compliant annotated material. This example corresponds to a Moodle site connected to the intelligent system.

The learner is visiting a KO, which is part of the concept container `Network Simulation` (the one corresponding to the code shown in Fig. 7 and the INTU-ITEL system is showing its content recommendations on the top-left side of the window. The preferred learning pathway of the student is the once labelled as `HasLogicalNextstep` thus two KOs belonging to the next CC in that pathway, CC `IPNetworking`, are recommended.

To conclude, we have shown that the default RDF Export function of Semantic MediaWiki, when the imported elements are used consistently, is able to preserve the pedagogical and formal semantics encoded in the PO. Therefore, those elements can be used by pedagogical teams in order to annotate the wiki pages of the course with pedagogical metadata. Those metadata will be the input for intelligent tutor systems like INTUITEL, which will recommend students learning pathways adapted to their particular needs and context.

6   Usability Test

An user evaluation study was conducted to study the usability and the efficiency of the proposed approach from a teacher or instructional point of view. The goal of the evaluation is to test, if the approach is useful for pedagogical teams while they create content and annotate content with pedagogical metadata while measuring the overall user satisfaction. Improvements in learning performance is not considered and tested since the proposed approach only addresses authoring workflows. Due to the fact that exported learning material can not be inconsistent because of the description logic family upon which Semantic MediaWiki's knowledge representation formalism is defined, we did not test for that in the evaluation.

If users are satisfied with the approach, we can assume that Semantic MediaWikis are a promising tool for the production and annotation of learning content with pedagogically meaningful and expressive ontological semantics. We assume that users are satisfied if the usability and the efficiency as rated by the users shows positive results. Since we additionally assume that sometimes online courses are created by people who are none computer experts, we compared users from departments of Computer Technology and users from departments of Education to research if the usability of SMW is acceptable for both user groups. Thus the research questions of our empirical study are:

RQ1: Does the usability test indicate an user satisfaction (usability and efficiency) in the positive half of the usability test, i.e., a value above 3.5?

RQ2: Does the measured usability show no differences in user satisfaction among faculties from Computer Technology departments and from departments of Education?

For testing the usability, we used a freely available standard usability test. Since we conducted a post-test, we decided to use the Post-Study System Usability Questionnaire (PSSUQ) as suggested by Lewis [42]. In this test, system usefulness, information quality and interface quality are measured. Since our research questions asks for the overall user satisfaction only, we used the test as a general indicator for usability. This is in line with the psychometric evaluation of the test [43].

The Hypotheses to be evaluated in this study are

H1: The usability test indicates an user satisfaction (usability and efficiency) in the positive half of the usability test result, i.e., a value above 3.5.

H2: The usability test indicates that there are no relevant differences in terms of user satisfaction among users from departments of Computer Technology and users from departments of Education.

For the first hypothesis, we used the middle of the scales as decision criterion. If the results are in the positive half of the scale, we consider an acceptable user satisfaction as indicated. For the second hypothesis, we used two decision criteria: If the results for both groups are in the positive half of the scale and do not show significant differences, we assume that the usability of SMW is acceptable for people from departments of Education as well as for people from departments of Computer Technology.

For the usability test, a Semantic MediaWiki was set up. Example content was created. The INTUITEL PO and SLOM were imported. Staff members from departments of Computer Technology and from departments of Education, who were involved in the INTUITEL project, were invited to participate as experimental subjects. Since the survey was taken anonymously, the exact position of the single experimental subject who actually participated in the study is not known. We invited junior and senior researchers and all of them participated. Thus we know that junior and senior researchers participated in the study. The experimental subjects were selected since they were familiar with the INTUITEL PO and SLOM and the Webdidactic concept. Thus we can assume that the process of learning the metadata system and the vocabulary did not influence the results of the usability test.

Six junior and senior researchers from departments of Computer Technology and nine junior and senior researchers from departments of Education participated in the usability test. Access data to the prepared Semantic MediaWiki and a usability questionnaire were sent to the participants. On the starting page of the Semantic MediaWiki, background information and instructions for the test were provided. The background information included (i) links to the imported ontologies, (ii) links to sample courses to illustrate structural coherences, and (iii) links to Semantic MediaWiki support pages. Thus participants could easily look up classes and properties, copy annotations from existing pages and retrieve information from available help pages. The instructions for the test as shown in Fig. 9 were provided on the starting page after the background information.

The usability test was provided as a spreadsheet file. The participating experimental subjects filled in the questionnaire in the spreadsheet and sent the file to the research team. The items were based on the usability test as suggested by Lewis [42]. The labels (SYS and INFO) were taken from Lewis and indicate two subscales of the usability test. Since our Hypotheses refer to the overall user satisfaction, the subscales are not relevant for our study. The items were adopted

```
2.2 Conduct the test

Task 1: Create individual terms, ie., annotations in form of properties and/or categories
that you will use for annotating the course you create in Task 2.

Task 2: Create a 9-page-course with 3 Concept Containers (CCs) and 3 Knowledge Objects (KOs)
per CC, where two learning pathways are used in each CC and among the CCs.
  1. Create the first page of your test course as a knowledge domain, i.e., add it to the
     category 'KnowledgeDomain' (edit this page to do this):
     Example 1
     Example 2
     YOUR FIRST PAGE HERE (copy the line above and edit it to create your first page)
  2. Create 3 exemplary concept containers (CCs).
  3. Create two macro learning pathways (MLP) between the CCs
  4. Create 3 exemplary knowledge objects (KOs) in each CC and create/copy example content.
  5. Annotate each KO. Use at least one annotation property for knowledge type,
     media type and level.
  6. Create two micro learning pathways in one of the CCs
```

Fig. 9. Excerpt of the task descriptions for the usability testing

in order to reflect the actual test situtation where a Semantic Media Wiki was used to annotate content with INTUITEL metadata.

In the usability test, a seven point Lickert scale was used with 7 as "strongly agree" and 1 as "strongly disagree". As statistical methods to analyze the observed results, we mainly used descriptive statistics. We calculated the mean for each item, the overall mean, and the means for faculties from departments of Computer Technology and from departments of Education.

For our first hypothesis ("*The usability test indicates an user satisfaction (usability and efficiency) in the positive half of the usability test result, i.e., a value above 4*") the results show that the overall mean of all answers was $\overline{x} = 4.45$. This is nearly 0.5 above the middle of the scale, while the result is in the lower part of the positive half of the scale. Additionally, the individual averages of most users who participated in the study were in the positive part of the scale, ranging from 3.79 to 5.64. Three users rated the usability slightly below the middle of the scale. Thus 80 % of the users rated the overall usability as positive. The mean of the items that ask for efficiency (SYS 3, SYS 5, SYS 6, SYS 7, SYS 8) is 4.74. The mean of all other items is 4.26. Both values are in the positive half of the scale. Thus, users feel that the usability and the efficiency of the system is good. These results do not falsify our hypothesis H1, which we thus keep up.

For our second hypothesis ("*The usability test indicates that there are no relevant differences in terms of user satisfaction among users from departments of Computer Technology and users from departments of Education.*") the results show that the average of faculties from departments of Computer Technology was $\overline{x} = 4.85$, while the average of faculties from departments of Education was $\overline{x} = 4.16$. Both results are clearly in the positive half of the scale. The difference appears small. To test the relevance of the difference, we applied the procedure as suggested by [44]. At first, a descriptive investigation for normal distribution has been conducted (according to Rasch et al., a Kolgomorov-Smirnow-Test is not appropriate for our sample size). The distribution has a slight skewness to the right, is relatively flat, and does not show relevant deviations from a normal distribution. Next, a Levene-Test for $H_0: \sigma_1^2 = \sigma_2^2$ has been calculated. The result is highly significant (Levene-test, $p < 0.001$). Thus, we have to assume that the variances are heteroscedastic. This has been considered in the degrees of freedom that have been used in the calculation of the t-test. Since we assume differences among researchers from departments of Computer Technology and researchers from department of Education, a t-test for $H_0: \overline{x}_1 = \overline{x}_2$ has been calculated. As we have no prior data that could justify a single sided test, we calculated a double sided test. The result shows that the difference is significant (t-test, $p < 0.01$). There is a significant difference among faculties from departments of Computer Technology and faculties from departments of Education. This result does falsify our hypothesis H2, which we thus reject.

With respect to H2, the difference between faculties from departments of Computer Technology and faculties from departments of Education could have been expected, since it is necessary to enter Semantic MediaWiki markup syntax into the content pages. An interface that supports the data entry process, for instance by drop down menus, or a syntax checking method was not provided. Since writing markup code is uncommon for faculties from departments of Education, it is not astonishing that the results are lower. In turn, it might be considered as astonishing, that the people from departments of Education still judged the system with an overall positive result. Additionally, the difference is significant, but not very high. Thus the results show the potential of using Semantic MediaWikis as a tool to create semantically enriched content for teaching and learning and indicate a need for usability improvements.

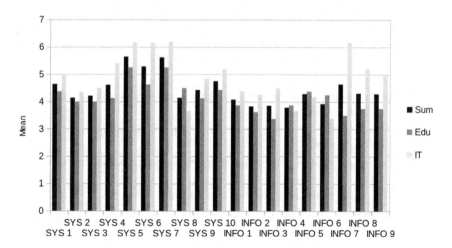

Fig. 10. Results of the usability test

The overall results as expressed in the overall means reported before are also clearly visible in the means per item as shown in Fig. 10. In the figure, the means per item for all users are shown as Sum. The x-axis refers to the items of the usability test that are listed in Table 3. All means for all items are above 3.5, with one exception. The item "It is easy to find the information I need (INFO 3)" was judged negatively by people from departments of Education. At the same time, the difference between the people from departments of Computer Technology and people from departments of Education was relatively high for this item. This can be explained by the fact the information about the metadata system was provided as import result pages in the Semantic MediaWiki. These

Table 3. Items of usability test

Label	Item
SYS 1	Overall, I am satisfied with how easy it is to use Semantic Media Wiki as an authoring tool
SYS 2	It was simple to use this system to create pedagogically meaningful annotations
SYS 3	I can effectively complete the given tasks
SYS 4	I am able to create individual terms and map them to the imported vocabulary terms
SYS 5	I am able to efficiently create new courses (Knowledge Domains)
SYS 6	I am able to efficiently create Concept Containers (CCs)
SYS 7	I am able to efficiently create Knowledge Objects (KOs)
SYS 8	I am able to efficiently create Learning Pathways ·
SYS 9	It was easy to learn how to use collaboratively created annotations using this system
SYS 10	I feel comfortable using this system
INFO 1	Whenever I make a mistake using this system, I recover easily and quickly
INFO 2	The information (such as online help, on-screen messages, and other documentation) provided with Semantic Media Wiki is clear
INFO 3	It is easy to find the information I needed
INFO	The information provided for using the imported pedagogical terms is easy to understand
INFO 5	The information provided is effective in helping me complete the tasks and scenarios
INFO 6	The organization of information using the provided vocabulary terms is clear
INFO 7	I feel comfortable using the Semantic Media Wiki syntax
INFO 8	This system has all the expressive features and capabilities I expect it to have
INFO 9	Overall, I am satisfied with this system

pages show the information like code with hypertext markup. Thus these pages were most probably difficult to read for people from departments of Education.

The assumption that working with syntax is unfamiliar for people from departments of Education and thus reduces the usability in terms of retrieving necessary information and entering metadata is supported by the answers on item INFO 7 "I feel comfortable using the Semantic Media Wiki Syntax". While the average of people from departments of Computer Technology was 6.17, the average of people from departments of Education was 3.5. This suggests that Semantic MediaWikis as used in our study, are appropriate for computer specialists, but should be enhanced with tools to enter metadata for non computer specialists. Still, the overall satisfaction which was asked for in item SYS 1 was quite high for all participants. And even the creation of individual terms, which is not an easy concept for users from departments of Education, was judged clearly positive. This again shows the high potential of Semantic MediaWikis as a tool to create semantically enriched content for teaching and learning.

7   Limitations

In this section, we critically review the presented approach and highlight its limitations. These discussions serve as indicators for directing future work and research in the given domain:

1. *No guarantee for schema compliance during the authoring process*
   SMW has been created with the idea in mind to extend wikis with machine-processability of their content to suport knowledge organization and sharing while maintaining its inherent wiki authoring style [7,23]. The presented approach, therefore, can not support checking for schema compliance during authoring processes per default—in particular due to the following reasons:
   (i) SMW does not make any assumptions regarding existing or prescribing schemas; therefore, it is not possible to define a schema against which created learning content can be checked for compliance.
   (ii) The knowledge representation formalism underlying SMW is built upon a description logic fragment that does not allow for the creation of unsatisfiable assertions, i.e., the data created with SMW can per design never be inconsistent (cf. limitation #2).
   These facts also limit the model checking capabilities of SMW and the automated transformation of learning content into other formats such as IMS Learning Design and SCORM (see Sect. 2.3). Such transformations still require a priori specification of mapping declarations plus external tools or activities for building content packages. Checks whether the imported elements are used correctly and annotations are consistent with PO and SLOM semantics can only be conducted posterior to an authoring process when learning material is exported and requires the deployment of an external OWL-DL-compliant reasoner[21]. An integration of consistency and compliance

---

[21] One possibility to overcome this limitation is integrating DL reasoning capabilities via a Semantic MediaWiki Extension.

checks in the authoring process, e.g., in form of an Extension, would have a positive impact on INFO 1 and INFO 2 (see Fig. 10).

2. *Limited expressivity of Semantic MediaWiki's knowledge representation formalism*
   Semantic MediaWiki's knowledge representation formalism is a less expressive subset of the description logic upon which OWL and hence the PO and SLOM are defined (cf. Sect. 4.1). Due to this unilateral incompatibility, only a limited set of the TBox and RBox axioms, in particular constituting axioms, can be imported but not axioms that use OWL DL language elements for defining the semantics of those terms (e.g., disjointness, quantifiers, inverse relationships, equivalence, or class membership restrictions etc.). This prevents SMW content authors to utilize the entire feature set of pedagogical semantics encoded in the PO and SLOM during the authoring process. While it is possible to use imported terms as intended, i.e., according to the semantics defined in the PO, SMW's default reasoning capabilities do not allow for consistency or compliance checking during the authoring process, as indicated by the comparatively low scores of INFO 1 and INFO 2.

3. *Creation of self-contained learning units requires external tools such as the INTUITEL Merger*[22]. The presented approach, in its current version, exports the articles' content plus contained annotations in form of RDF graphs and requires additional tools for building self-contained learning content packages that can be used in an INTUITEL-enabled Learning Management System (LMS)[23]. If concrete media files such as PDF documents, presentations, video or audio files are linked in exported article pages, only their dereferencable URIs are exported but not their actual content. In order to retrieve such media and amalgamate it with wiki content and corresponding annotations, their URLs need to be dereferenced and its media content stored locally to build a self-contained content package. This involves a remapping of multimedia content file URIs in the exported RDF graphs since they still point to the URI of the Semantic MediaWiki that hosts those files. The INTUITEL Merger has been implemented as a self-contained tool that performs URI dereferencing and remaps the URIs of contained content files to their location in a SLOM content package, i.e., the learning unit loaded into a LMS. Future work aims at integrating the INTUITEL Merger with SMW in order to create self-contained content packages directly within SWM.

4. *Import of ABox data (instances) by default requires manual intervention*
   ABox data can not be imported directly by default and need to be added manually to a Semantic MediaWiki system. This has some implications for ontologies that also contain instance data. The PO, for instance, defines one specific instance for each knowledge and media type to enable the formulation of assertions such that a KO participates in a `hasMediaType`-relation

---

[22] The INTUITEL Merger is specified in Deliverable 6.1: http://www.intuitel.de/wp-content/uploads/2015/06/INTUITEL_318496_D6_1_MergerDoc.pdf.

[23] While this aspect has no direct influence on the authoring process and its usability, which was studied in the evaluation, it might impede market penetration of the presented approach specifically and SWMs in generally.

to the singleton instance that corresponds to the specific media type class[24]. Although the model-theoretic semantic of such an assertion is different, it allows for the usage of instances both on ABox and TBox level and corresponds to the notion of *DL nominals* (see [33–35]). This concept is useful in situations where classes should also be used as single individuals and where it seems unnatural to have multiple instances of one class (cf. [45]). To overcome this limitation, we have created an ontology import tool for Semantic MediaWiki[25] that analyzes all ABox axioms of an ontology, creates article pages of constituting elements, and assigns them to the corresponding categories, i.e., the classes of which the instances are members of in the ontology. It also analyzes relationships between instances and tries to resemble them in the target SWM system.

5. *Annotating content requires knowledge of the vocabulary and the wiki markup language*
   In the current version, the system requires sound knowledge of the metadata vocabulary in the PO that is used to annotate content. Our usability study showed that this might keep users from using Semantic MediaWiki as an editing tool. This seems similar to editing content itself, which required learning the wiki markup language and their formal semantics until recently. The necessity to learn the wiki markup language was perceived as a restriction by non-computer specialists that kept many people away from contributing to Wikipedias. To overcome this restriction, the VisualEditor for MediaWiki has been developed and deployed in most Wikipedias. This makes the production of content for non-computer specialists much easier and suggests to develop an enhancement for Semantic MediaWiki that supports the annotation of content with a tool that might for instance be based on drop down lists.

The last point in particular illustrates that the presented approach requires an additional technical annotation facility that not only helps non-technical experts in understanding the intended semantics of imported terms but also shows additional information about them, i.e., how they are linked together. One possibility to do that is by defining semantic templates[26] specifically designed for imported ontology elements. Such templates can then be used in combination with Semantic Forms[27] to guide non-technical users in particular in the annotation process and hence contribute towards an improved usability of the system.

---

[24] The same applies for knowledge types.

[25] The Ontology Import Tool will be released as open source and can be downloaded from the official INTUITEL project Web site as well as from the Open Source section of the FZI Research Center for Information Technology Web site https://www.fzi.de/forschung/open-source/.

[26] http://semantic-mediawiki.org/wiki/Help:Semantic_templates.

[27] https://www.mediawiki.org/wiki/Extension:Semantic_Forms.

Other limitations, that are not related to the primary focus of this work but might be required by learning content authors are, e.g., true synchronous collaborative editing of wiki articles and real-time change tracking (see e.g. [11]).

8 Conclusion

In this work, we present an approach for the collaborative annotation of learning material using pedagogically well-defined semantics in a Semantic MediaWiki system. We introduce Web Didactics as knowledge organization system together with the Pedagogical Ontology and the Semantic Learning Object Model as manifestations of the Web Didactics concepts. We also discuss limitations of Semantic MediaWiki's knowledge representation formalism and demonstrate how pedagogically meaningful terms from the PO and SLOM can be made available in a Semantic MediaWiki system through import and mapping declarations. These declarations constitute the foundation to map imported ontology terms to the individual vocabulary used in a Semantic MediaWiki for the annotation of learning content so that content developers are not forced to adapt to new vocabularies but can use the terms and classification systems they are familiar with. Through a network design course taught at the University of Valladolid, we show how imported terms can be used for the annotation of real course material and how the inherently defined pedagogical semantics can be preserved when the course content is exported as Linked Data. This use case also demonstrates that when PO and SLOM terms are used in a consistent manner, a direct mapping between those terms and Semantic MediaWiki language elements can be realized without compromising the formal semantics of the PO and SLOM. We study the impact of the presented Semantic MediaWiki-based annotation approach in terms of its usability for researchers and lectures from Computer Technology and Education departments. Results were promising and proved a good overall usability. This reveals the potential of Semantic MediaWikis to create semantically enriched content for teaching and learning.

Acknowledgements. Parts of this work are compiled from Deliverables 4.1 and 4.2 created within the INTUITEL project, which is financed by the European Commission under Grant Agreement No. 318496.

References

1. Polsani, P.R.: Use and abuse of reusable learning objects. J. Digit. Inf. **3** (2003)
2. Dodero, J.M., Díaz, P., Sarasa, A., Sarasa, I.: Integrating ontologies into the collaborative authoring of learning objects. J. Univ. Comput. Sci. **11**, 1568–1575 (2005)
3. Sosnovsky, S., Yudelson, M., Brusilovsky, P.: Community-oriented course authoring to support topic-based student modeling. In: Dicheva, D., Mizoguchi, R., Capuano, N., Harrer, A. (eds.) Proceedings of the Fifth International Workshop on Ontologies and Semantic Web for E-Learning (SW-EL 2007) at AIED 2007, Marina Del Ray, CA, USA, pp. 91–100, July 2007

4. Brut, M.M., Sedes, F., Dumitrescu, S.D.: A semantic-oriented approach for organizing and developing annotation for e-learning. IEEE Trans. Learn. Technol. 4(3), 239–248 (2011)
5. Bratsas, C., Dimou, A., Parapontis, I., Antoniou, I., Alexiadis, G., Chrysou, D.-E., Kavargyris, K., Bamidis, P.: Educational semantic wikis in the linked data age: the case of msc web science program at aristotle university of thessaloniki. In: Dietze, S., d'Aquin, M., Gasevic, D., Sicilia, M.-A. (eds.) Proceedings of Linked Learning: The 1st International Workshop on eLearning Approaches for the Linked Data Age, co-located with the 8th Extended Semantic Web Conference, ESWC 2011 (2011)
6. Pahl, C., Holohan, E.: Applications of semantic web technology to support learning content development. Interdisc. J. E-Learn. Learn. Objects 5, 1–25 (2009)
7. Li, Y., Dong, M., Huang, R.: Designing collaborative e-learning environments based upon semantic wiki: from design models to application scenarios. Educ. Technol. Soc. 14(4), 49–63 (2011)
8. Boyce, S., Pahl, C.: Developing domain ontologies for course content. Educ. Technol. Soc. 10(3), 275–288 (2007)
9. Protégé: The Protégé Ontology Editor and Knowledge Acquisition System 2007. http://protege.stanford.edu/. Accessed 25 Jan 2007
10. Holohan, E., Melia, M., Mcmullen, D., Pahl, C.: Adaptive e-learning content generation based on semantic web technology. In: Workshop on Applications of Semantic Web Technologies for e-Learning, SW-EL@ AIED 2005, 18 July 2005
11. Rahhal, C., Skaf-Molli, H., Molli, P., Weiss, S.: Multi-synchronous collaborative semantic wikis. In: Vossen, G., Long, D.D.E., Yu, J.X. (eds.) WISE 2009. LNCS, vol. 5802, pp. 115–129. Springer, Heidelberg (2009)
12. Swertz, C.: Web-Didaktik. eine didaktische Ontologie in der Praxis. Medienpädagogik 2(4) (2005)
13. Meder, N.: Web-Didaktik. Eine neue Didaktik webbasierten, vernetzten Lernens. Bertelsmann, Bielefeld, January 2005
14. Meder, N.: Der Lernprozess als performante Korrelation von Einzelnem und kultureller Welt. eine bildungstheoretische Explikation des Begriffs. Spektrum Freizeit, no. 1/2, pp. 119–135 (2007)
15. Swertz, C., Barberi, A., Forstner, A., Schmöz, A., Henning, P., Heberle, F.: A sms with a kiss. towards a pedagogical design of a metadata system for adaptive learning pathways. In: Proceedings of World Conference on Educational Media and Technology 2014, Tampere, Finland, pp. 991–1000, Association for the Advancement of Computing in Education (AACE), June 2014
16. Swertz, C.: Didaktisches Design. ein Leitfaden für den Aufbau hypermedialer Lernsysteme mit der Web-Didaktik (2004)
17. Herbart, J., Wendt, H.: Umriss pädagogischer Vorlesungen. Reclams Universal-Bibliothek, Reclam (1890)
18. Oevermann, U.: Theoretische Skizze einer revidierten Theorie professionalisierten Handelns. In: Combe, A., Helsper, W. (eds.) Pädagogische Professionalität. Untersuchungen zum Typus pädagogischen Handeln, no. 1230 in stw, (Frankfurt a.M.), pp. 70–182, Suhrkampp (1996)
19. Koper, R., Miao, Y.: Using the ims ld standard to describe learning designs (2007)
20. Meder, N.: L3-ein didaktisches Modell als Impulsgeber. In: Ehlers, U.-D., Gerteis, W., Holer, T., Jung, H.W. (Hrsg.) E-Learning Services im Spannungsfeld von Pädagogik, Ökonomie und Technologie (2003)
21. Meder, N.: Web-Didaktik: eine neue Didaktik webbasierten, vernetzten Lernens. Wissen und Bildung im Internet, Bertelsmann (2006)

22. Introduction to semantic mediawiki. http://www.semantic-mediawiki.org/wiki/ Help:Introduction_to_Semantic_MediaWiki Mai 2009. Stand 13.5.2009
23. Krötzsch, M., Vrandečić, D., Völkel, M.: Semantic MediaWiki. In: Cruz, I., Decker, S., Allemang, D., Preist, C., Schwabe, D., Mika, P., Uschold, M., Aroyo, L.M. (eds.) ISWC 2006. LNCS, vol. 4273, pp. 935–942. Springer, Heidelberg (2006)
24. Vrandecic, D., Krötzsch, M.: Reusing ontological background knowledge in semantic wikis. In: Völkel, M., Schaffert, S. (eds.) Proceedings of the First Workshop on Semantic Wikis - From Wiki To Semantics, Workshop on Semantic Wikis, AIFB, ESWC 2006, June 2006
25. Bizer, C., Heath, T., Berners-Lee, T.: Linked data - the story so far. Int. J. Semantic Web Inf. Syst. 5(3), 1–22 (2009)
26. Dean, M., Schreiber, G.: Owl web ontology language reference. w3c recommendation, Web Ontology Working Group, World Wide Web Consortium (2004)
27. Hitzler, P., Krötzsch, M., Rudolph, S.: Foundations of Semantic Web Technologies. CRC Press, Boca Raton (2010)
28. d'Aquin, M.: Linked data for open and distance learning. Commonwealth of Learning (COL) (2012)
29. Krieger, K., Rösner, D.: Linked data in e-learning: a survey. Semantic Web J. (1), 1–9 (2011)
30. Inaba, A., Mizoguchi, R.: Learning design palette: an ontology-aware authoring system for learning design. In: Proceedings of the International Conference on Computers in Education (2004)
31. Duval, E.: Learning object metadata (LOM) Technical report, 1484.12.1 - 2002. IEEE (2002)
32. Aroyo, L., Mizoguchi, R., Tzolov, C.: Ontoaims: Ontological approach to courseware authoring. In: Mitchell, K., Lee, K. (eds.) The "Second Wave" of ICT in Education: From Facilitating Teaching and Learning to Engendering Education Reform, International Conference on Computers in Education 2003, pp. 1011–1014. AACE, December 2004
33. Baader, F., Calvanese, D., McGuinness, D., Nardi, D., Patel-Schneider, P.: The Description Logic Handbook: Theory, Implementation and Applications. Cambridge University Press, Cambridge (2003)
34. Rudolph, S.: Foundations of description logics. In: Polleres, A., d'Amato, C., Arenas, M., Handschuh, S., Kroner, P., Ossowski, S., Patel-Schneider, P. (eds.) Reasoning Web 2011. LNCS, vol. 6848, pp. 76–136. Springer, Heidelberg (2011)
35. Krötzsch, M., Simančík, F., Horrocks, I.: Description logics. IEEE Intell. Syst. 29, 12–19 (2014)
36. Guarino, N., Oberle, D., Staab, S.: What is an ontology? In: Handbook on Ontologies, pp. 1–17. Springer, Heidelberg (2009)
37. Stuckenschmidt, H.: Ontologien: Konzepte, Technologien und Anwendungen. Informatik im Fokus, 2nd edn. Springer, Heidelberg (2011)
38. Owl 2 web ontology language document overview. W3C Recommendation, October 2009
39. W3C, Owl 2 web ontology language profiles, 2nd edn., November 2012. http://www.w3.org/TR/owl2-profiles/
40. Schneider, M.: OWL 2 web ontology language RDF-based semantics, 2nd edn., December 2012
41. Jacobs, I., Walsh, N.: Architecture of the world wide web, vol. 1. http://www.w3.org/TR/webarch/
42. Lewis, J.R.: Ibm computer usability satisfaction questionnaires: psychometric evaluation and instructions for use. Int. J.Hum.Comput. Interact. 7(1), 57–78 (1995)

43. Lewis, J.R.: Psychometric evaluation of the post-study system usability questionnaire: the pssuq. In: Proceedings of the Human Factors and Ergonomics Society Annual Meeting, vol. 36(16), pp. 1259–1260 (1992)
44. Rasch, B., Friese, M., Hofmann, W., Naumann, E.: Quantitative Methoden Band 1: Einführung in die Statistik für Psychologen und Sozialwissenschaftler, 4. Springer (2014)
45. Zander, S., Awad, R.: Expressing and reasoning on features of robot-centric workplaces using ontological semantics. In: IEEE/RSJ International Conference on Intelligent Robots and Systems (2015) (to be published)

Exploiting Semantics
in Collaborative Spaces

Discovering Wikipedia Conventions Using DBpedia Properties

Diego Torres[1]([✉]), Hala Skaf-Molli[2], Pascal Molli[2], and Alicia Díaz[1]

[1] LIFIA, Fac. Informática,
Universidad Nacional de La Plata, 1900 La Plata, Argentina
{diego.torres,alicia.diaz}@lifia.info.unlp.edu.ar
[2] Nantes University, LINA, 2, Rue de la Houssiniere, 44322 Nantes, France
{Hala.Skaf,Pascal.Molli}@univ-nantes.fr

Abstract. Wikipedia is a public and universal encyclopedia where contributors edit articles collaboratively. Wikipedia infoboxes and categories have been used by semantic technologies to create DBpedia, a knowledge base that semantically describes Wikipedia content and makes it publicly available on the Web. Semantic descriptions of DBpedia can be exploited not only for data retrieval, but also for identifying missing *navigational paths* in Wikipedia. Existing approaches have demonstrated that missing navigational paths are useful for the Wikipedia community, but their injection has to respect the Wikipedia convention. In this paper, we present a collaborative recommender system approach named BlueFinder, to enhance Wikipedia content with DBpedia properties. BlueFinder implements a supervised learning algorithm to predict the Wikipedia conventions used to represent similar connected pairs of articles; these predictions are used to recommend the best convention(s) to connect disconnected articles. We report on an exhaustive evaluation that shows three remarkable elements: (1) The evidence of a relevant information gap between DBpedia and Wikipedia; (2) Behavior and accuracy of the BlueFinder algorithm; and (3) Differences in Wikipedia conventions according to the specificity of the involved articles. BlueFinder assists Wikipedia contributors to add missing relations between articles, and consequently, it improves Wikipedia content.

Keywords: Semantic Web · Social web · DBpedia · Wikipedia · Collaborative recommender systems

1 Introduction

Semantic Web technologies facilitate search and navigation on the Web, while they can be additionally used to extract data from the Social Web, e.g., DBpedia is built with data extracted from Wikipedia[1] infoboxes and categories. On the other hand, knowledge managed and encoded by Semantic Web technologies can

---

[1] http://www.wikipedia.org.

© Springer International Publishing Switzerland 2016
P. Molli et al. (Eds.): SWCS 2013/2014, LNCS 9507, pp. 115–144, 2016.
DOI: 10.1007/978-3-319-32667-2_6

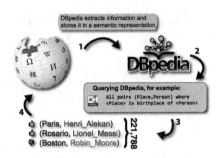

(a) Information flow from Wikipedia to DBpedia

(b) Complete cycle of information flow between Wikipedia and DBpedia

Fig. 1. Information flow between Social Web and Semantic Web

enhance data published in the Social Web; for instance, semantic annotations of data have been used to improve the Facebook graph search [1]. However, to the best of our knowledge, DBpedia [2] has not been exploited to improve Wikipedia. In this paper, we propose BlueFinder an approach to enhance the content of Wikipedia with data inferred in DBpedia.

Although Wikipedia is collaboratively edited by large user communities, Wikipedia links are not semantically defined, e.g., no types are attached to these links. In the context of Semantic Web, Wikipedia links are translated into properties in DBpedia, and they are semantically described using RDF vocabularies, i.e., DBpedia encodes semantics that is not represented in Wikipedia and provides a more expressive representation of Wikipedia links. Therefore, DBpedia allows for retrieving information that is not available in Wikipedia [3]. To illustrate, Listing 1.2 presents a SPARQL [4] query named $Q1$ to retrieve people and their born place[2] using `db-prop:birthplace`. Nevertheless, if $Q1$ is executed against the DBpedia endpoint[3], the answer includes more people than those obtained by navigating from the Wikipedia place article. The evaluation of query $Q1$ retrieves 409,812 (place, person) pairs from the DBpedia endpoint. Meanwhile, if we navigate from places to people in Wikipedia, we only obtain 221,788 connected pairs. Two Wikipedia articles are connected if a regular Wikipedia user can navigate from one article to another through a navigational path. A navigational path with a length larger than five is unreachable by a regular user [5–7]; so those articles are considered as disconnected. Thus, only 54 % of places in Wikipedia have a navigational path to those people who were born there. In this paper, we aim at adding missing navigational paths in Wikipedia, and enhancing Wikipedia content. Thus, we contribute to complete the virtuous cycle of information flow between Wikipedia and DBpedia as illustrated in Fig. 1b.

To measure how important the gap between Wikipedia and DBpedia, we choose the most popular classes of DBpedia defined in [2], and the properties

---

[2] A place could be a Country, Province, City, or State.

[3] DBpedia of July 2013.

with the highest number of triples. We call these properties *relevant properties*. Listing 1.1 shows the SPARQL query that retrieves the relevant properties that relate instances of the classes `db-o:Person` and `db-o:Place`.

```
prefix db-o:<http://dbpedia.org/ontology/>
select ?p (count(distinct ?o) as ?count)
where {    ?s ?p ?o.
           ?s rdf:type    db-o:Person.
           ?o rdf:type    db-o:Place
}
group by ?p
order by ?count
```

Listing 1.1. SPARQL query to retrieve relevant properties that relate instances of the classes `db-o:Person` and `db-o:Place`

We observe the same phenomenon when querying DBpedia using *relevant properties* of other classes, e.g., `db-o:Person`, `db-o:Place`, or `db-o:Work`, as shown in Table 1. The last two columns of the table provide the number of connected pairs obtained by a SPARQL query and the amount of disconnected pairs in Wikipedia for a specific property, respectively.

Some connected resources in DBpedia are disconnected in their corresponding Wikipedia articles, i.e., resources can be navigated in DBpedia while it is

Table 1. Results of 20 SPARQL queries for 20 properties and different classes.

DBpedia property	From class	To class	# DBpedia connected pairs	# Wikipedia disconnected pairs
$prop_1$: birthPlace	Place	Person	409, 812	221, 788
$prop_2$: deathPlace	Place	Person	108, 148	69, 737
$prop_3$: party	PoliticalParty	Person	31, 371	15, 636
$prop_4$: firstAppearance	Work	Person	1, 701	142
$prop_5$: recordLabel	Company	Person	25, 350	14, 661
$prop_6$: associatedBand	MusicalWork	Person	365	73
$prop_7$: Company	Software	developer	14, 788	2, 329
$prop_8$: recordedIn	PopulatedPlace	MusicalWork	28, 351	27, 896
$prop_9$: debutstadium	Building	Athlete	595	393
$prop_{10}$: producer	Artist	MusicalWork	70, 272	32, 107
$prop_{11}$: training	Building	Artist	171	109
$prop_{12}$: previousWork	Album	MusicalWork	72, 498	3, 887
$prop_{13}$: recordLabel	Company	MusicalWork	118, 028	75, 329
$prop_{14}$: starring	Person	Film	164, 073	42, 584
$prop_{15}$: country	PopulatedPlace	Book	19, 224	17, 281
$prop_{16}$: city	PopulatedPlace	Educational institution	34, 061	8, 681
$prop_{17}$: associatedBand	Band	MusicalArtist	24, 846	4, 100
$prop_{18}$: fromAlbum	Album	Single	18, 439	1, 268
$prop_{19}$: location	PopulatedPlace	Airport	10, 049	2, 660
$prop_{20}$: notableWork	Book	Person	1, 510	73

not possible to navigate equivalent resources in Wikipedia. We call this missing navigational paths *information gap* between Wikipedia and DBpedia. To illustrate the gap, we analyzed 1,153,652 connected pairs in DBpedia, and we found that 540,434 pairs were disconnected in Wikipedia. Consequently, the value of the information gap between Wikipedia and DBpedia is important. Figure 2 details the number of information gap between Wikipedia and DBpedia for the properties detailed in Table 1. In order to evaluate the usefulness of adding these navigational paths, we carry out a social evaluation [8]. In this evaluation, we have manually added missing navigational paths for 211 disconnected pairs and after one month, we analyzed how many navigational pairs were accepted and how many were rejected by the Wikipedia community. As detailed in [8], 90 % of new navigational paths were accepted and 10 % were rejected. Although the rejected navigational paths had respected the semantics of the relation, they were more general than those used by the community. For example, the proposed navigational path to connect *(Edinburgh, Charlie Aitken)*[4] with the DBpedia property *is birthplace of* was `Edinburgh / Category:Edinburgh / Category:People_from_Edinburgh / Charlie_Aitken`[5]. Wikipedia community argued that the category *People from Edinburgh* is too general and the more specific category *Sportspeople from Edinburgh* is a more appropriate link.

For adding missing navigational paths, it is mandatory to study how the Wikipedia articles are connected respecting the Wikipedia conventions[6].

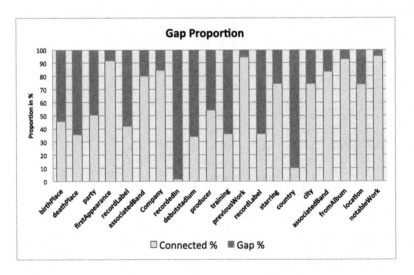

Fig. 2. Gap proportion for the twenty DBpedia properties of Table 1

---

[4] Charlie Aitken (footballer born 1942).

[5] It must be read as *from Edinburgh article, the user navigates through a link to the category Edinburgh then he or she navigates to People from Edinburgh category, and then to Charlie Aitken article.*

[6] http://en.wikipedia.org/wiki/Wikipedia:Conventions.

Wikipedia community has defined conventions that cover a wide diversity of topics: writing style, context of the articles and relations among articles. *Categories*, *List of* pages, and *Navigation templates* are the conventions to design the navigation for one-to-many relations among Wikipedia articles. Additionally, the conventions could be defined by the community according to the specificity of the articles [8]. For example, the DBpedia property *is birthPlace of* that relates *Boston*[7] and *Tim Barsky*[8] is represented in Wikipedia by the navigational path `Boston / Category:Boston / Category:People_from_Boston / Tim_Barsky`. However, in case of Boston and *Donna Summer*[9], the same DBpedia property is represented by the navigational path `Boston / Category:Boston / Category: People_from_Boston / Category:Musicians_from_Boston / Donna_Summer`. Donna Summer is a musician from Boston and most of the Boston's musicians belong to the Category:Musicians_from_Boston. These differences in the convention used to express the birthplace of a person trigger the following new question: *How to find the Wikipedia convention for a navigational path?*

In this paper, we address the problem of finding Wikipedia convention(s) that represent a DBpedia property between pairs of Wikipedia articles. Then, it is possible to connect pairs of Wikipedia articles that used to be disconnected. Consequently, Wikipedia content will be improved. For example, according to DBpedia (and also Wikipedia), Boston is the birthplace of Robin Moore. However, a navigational path from Boston to Robin Moore does not exist in Wikipedia but does exist in DBpedia. Therefore, a user could ask *Which is the convention to represent the "is birthPlace of" relation for (Boston, Robin Moore) in Wikipedia?*

We introduce BlueFinder, a collaborative recommender system that recommends navigational paths that represent a DBpedia property in Wikipedia. BlueFinder pays special attention to the specificity of the resource types in DBpedia. It learns from those similar pairs already connected by Wikipedia community and proposes a set of recommendations to connect a pair of disconnected articles. BlueFinder recommender system presented in this paper is an optimization of the previous version published in [9].

Summary of Our Contributions:

1. We measure the information gap between DBpedia and Wikipedia for a set of twenty representative DBpedia properties.
2. We re-design and propose several optimizations for BlueFinder algorithm.
3. We propose a new Semantic Pair Similarity Distance (SPSD) function to measure the similarity between pairs of related articles based on the DBpedia types description.
4. We conduct an empirical evaluation that measures Precision, Recall, F1, Hit-rate, and the confidence of BlueFinder recommendations over the twenty properties of DBpedia. The results demonstrate that BlueFinder is able to fix in average 89 % of the disconnected pairs with good accuracy and confidence.

---

[7] http://en.wikipedia.org/wiki/Boston.

[8] http://en.wikipedia.org/wiki/Tim_Barsky.

[9] http://en.wikipedia.org/wiki/Donna_Summer.

The paper is organized as follows. Section 2 presents related works in the field of discovering and recommending links. Section 2 describes basic definitions used in this work. Section 4 describes the problem statement. Section 5 presents the BlueFinder approach and the algorithm description. An exhaustive evaluation is described in Sect. 6. Finally, conclusions and future works are presented in Sect. 7.

2   Related Work

Want et al. [10] introduce a *collaborative approach* to recommend categories to Wikipedia Articles. The approach consists of a two-step model. The first step collects similar articles to the uncategorized one in terms of incoming and out coming links, headings and templates. The second step lies on ranking the categories obtained by the related articles and selecting the best ranked. BlueFinder uses categorization but in another context. Categorization is used to express properties of DBpedia. Less related to our approach but in line with combining recommender systems and DBpedia, *MORE* [11] is a recommender system that uses DBpedia to recommend movies in a Facebook application. A Vector space model is used to compute semantic similarity. However, *MORE* uses DBpedia as a source data set to base the recommendations and not to improve Wikipedia. Panchenko et al. [12] propose to extract semantic relations between concepts in Wikipedia applying kNN algorithms called Serelex. Serelex receives a set of concepts and returns sets where articles are semantically related according to Wikipedia information and using Cosine and Gloss overlap distance functions. In addition to the lack of using DBpedia as semantic base, Serelex cannot describe the way that two concepts are related in Wikipedia according to a property. Singer et al. [13] compute semantic relatedness in Wikipedia by means of human navigational path. They analyze human navigation paths in order to detect semantic relatedness between Wikipedia articles. The main difference with BlueFinder is the absence of DBpedia as a support to describe semantic relatedness between Wikipedia concepts, as is exploited in our work with the Semantic Pair Similarity Distance. Di Noia et al. [14] introduce a strategy to find similarity among RDF resources in the Linked Open Data. Di Noia et al. present a content-based recommender system approach based on the Web of Data. As in BlueFinder, the similarity between resources is computed by means of semantic relationships. However, in [14] it is mandatory to discover the semantic relation among the resources and then to analyze a potential similarity. In BlueFinder, we already know that the two resources in each pair are related by the same property and then we only have to compare the types of description. Finally, the Di Noia approach is applied in the Web of Data world, and BlueFinder we complement and augment the information of the Social Web with information from the Web of Data.

Nunes et al. [15] introduce a recommender system to discover semantic relationships among resources from the Social Web. The work presents an approach to measure the connectivity between resources in a particular data set based on semantic connectivity store and co-occurence-based measure. The first metric is based on graph relations among entities; the second one relies on an approximation of the number of existing Web pages that contain the same labels in their body. Although the approach of Nunes et al. is closely related to BuleFinder, the main difference is the direction of the information flow. In Nunes et al. work, the information of the Social Web is used to improve the Semantic Web and not in the opposite direction as in BlueFinder.

Other works [16,17] aim at fixing missing direct links in Wikipedia, while [18] proposes an approach to complete Wikipedia infobox links with information extracted from Wikipedia by using Kylin. Works in [16–18] do not use Semantic Web features as BlueFinder does, moreover, BlueFinder fixes navigation paths rather than only direct links.

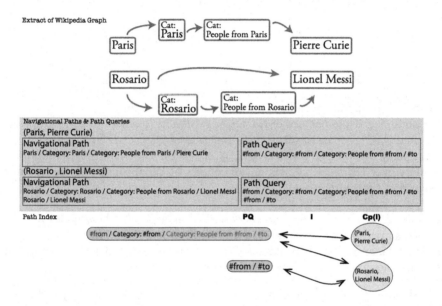

Fig. 3. Example of Wikipedia graph, navigational path, path queries and path index (Colour figure online)

DBpedia has not been exploited before to improve the content of Wikipedia. BlueFinder proposes to enhance the content of Wikipedia with data inferred in DBpedia, and to complete the cycle of information flow between Wikipedia and DBpedia as illustrated in Fig. 1b.

3   Preliminaries Definitions

DBpedia Knowledge Base: DBpedia knowledge base is a set of RDF triples built from data extracted from Wikipedia [2]. This knowledge base has a set of general properties and a set of infobox-specific properties, if the corresponding Wikipedia article contains an infobox. Each resource in DBpedia has types definition (*rdf:type*) coming from DBpedia ontology and Yago [19] ontology. The *foaf:isPrimaryTopicOf* property relates a DBpedia resource with its corresponding Wikipedia page. DBpedia provides a SPARQL endpoint to query DBpedia knowledge base. For example, the SPARQL query $Q1$ in Listing 1.2 retrieves from DBpedia the set of all pairs of Wikipedia pages (*from, to*) that are related by the DBpedia property *birthplace*.

```
prefix  db-o:<http://dbpedia.org/ontology/>
prefix  db-p:<http://dbpedia.org/property/>
prefix  foaf:<http://xmlns.com/foaf/0.1/>
select  ?fr  ?to
where   {    ?db_from  rdf:type  db-o:Person.
             ?db_to  rdf:type  db-o:Place.
             ?db_from  db-p:birthplace  ?db_to.
             ?db_from  foaf:isPrimaryTopicOf  ?fr.
             ?db_to  foaf:isPrimaryTopicOf  ?to
        }
```

Listing 1.2. Q1: DBpedia query for birthplace property.

For each property in Table 1, we define a corresponding SPARQL query Q_p that contains a triple pattern of the form: *?s db-p:p ?o* where p is the property. We define $Q_p(D)$ as the result of the evaluation of Q_p over D. In Listing 1.2, for the property $db - p : birthplace$, $Q1_{birthplace}(D)$ is the result of evaluation of the query over DBpedia, i.e., the set of couples of Wikipedia pages that relates persons and their birthplace.

Wikipedia Model: Wikipedia can be described as a graph where nodes are the Wikipedia articles (regular articles and categories) and hyperlinks are the edges. For example, an extract of Wikipedia graph is represented at the top of Fig. 3: boxes are the nodes and arrows are the edges.

Definition 1 (Wikipedia Graph). *$G = (W, E)$ where W is a set of nodes and $E \subseteq W \times W$ is a set of edges. Nodes are Wikipedia articles (wiki pages) and edges are links between articles. Given $w_1, w_2 \in W$, $(w_1, w_2) \in E$ if and only if there is a link from w_1 to w_2.*

Definition 2 (Navigational Path). *A navigational path $P(w_1, w_n)$ between two Wikipedia articles is a sequence of pages $w_1/ \ldots /w_n$, s.t. $\forall i \, w_i \in W \wedge \forall i, j : 1 \leqslant i < j \leqslant n$, $w_i \neq w_j$, $\forall i : 1 \leqslant i \leqslant n - 1$ where $(w_i, w_{i+1}) \in E$ is a link between w_i and w_{i+1}. w_1 and w_n are called the* source *page and the* target *page respectively. The length of a navigational path is the number of articles in the sequence, length $P(w_1, w_n) = n$.*

Definition 3 (Wikipedia Connected Pairs). *Let $Q_p(D)$ denote the result of the execution of the query Q_p against the dataset D. The set of pairs (f, t) in Wikipedia which are connected by a navigational path with length up to l, where $(f, t) \in Q_p(D)$ is defined as:*
$C_p(l) = \{(f, t) \in Q_p(D) \text{ such that } \exists P(f, t) \text{ and } length(P(f, t)) <= l\}.$

A path query is a generalization of similar navigational paths. Usually, regular expressions are used for expressing path queries [20]. Many works have been done on path queries in different domains [20–22]. We adapt the path query definition in [20,21] to the context of Wikipedia. We use regular expression patterns [20], i.e., patterns that include variables.

Definition 4 (Regular Expression Pattern [20]). *Let Σ be an alphabet, X be a set of variables, the set of regular expressions $R(\Sigma, X)$ over Σ can inductively defined by: (1) $\forall a \in \Sigma, a \in R(\Sigma, X)$. (2) $\forall x \in X, x \in R(\Sigma, X)$; (3) $\epsilon \in R(\Sigma, X)$. (4) If $\forall A \in R(\Sigma, X)$ and $\forall B \in R(\Sigma, X)$ then $A.B$, $A^* \in R(\Sigma, X)$; such that $A.B$ is the concatenation of A and B and A^* denotes the Kleene closure.*

Definition 5 (Language Defined by a Regular Expression Pattern [20]). *Let Σ be an alphabet, X be a set of variables, and $R, R' \in R(\Sigma, X)$ be two regular expression patterns. $L^*(R)$ is the set of words of $(\Sigma \bigcup X)^*$ defined by: (1) $L^*(\epsilon) = \{\epsilon\}$. (2) $L^*(a) = \{a\}$. (3) $L^*(x) = \Sigma \bigcup X$. (4) $L^*(R.R') = \{w'.w \mid w \in L^*(R)$ and $w' \in L^*(R')\}$. (5)$L^* (R^+) = \{w_1 \ldots w_k \mid \forall i \in [1...k], w_i \in L^*(R)$. (6) $L^*(R^*) = \{\epsilon\} \bigcup L^*(R+)$.*

A path query is a generalization of similar navigational paths by regular expressions patterns. A path query is defined by:

Definition 6 (Path Query). *A Wikipedia path query (in short path query) PQ $\in R(\Sigma, X)$ is a regular expression pattern. A pair of nodes (x, y) of G covers (or satisfies) a path query $PQ(x, y)$ over Σ and X if there exists a path P from x to y in G and a map μ from $\Sigma \bigcup X$ to $term(G)$ such that $\Lambda(P) \in L^*(\mu(R))$ where $\Lambda(P) = \Lambda(a_1) \ldots \Lambda(a_k)$ over $(\Sigma \bigcup X)^*$ is associated to the path $P = (a_1, ..., a_k)$ of G.*

In the context of Wikipedia $\Sigma = W$. For the purpose of this work, we limit X to two variables $X = \{\#from, \#to\}$. For a pair of Wikipedia pages (x, y) then, $\Lambda(\#\text{from}) = x$, $\Lambda(\#\text{to}) = y$ and $\Lambda(w) = w', w' \in W$, in case w includes in its a literal occurrence of the symbol #from or #to they will replace by x or y respectively, otherwise $w = w'$ (for example, $\Lambda(Category : \#from) = Category : Paris$ for the pair (Paris, Pierre Curie). Given a $Q_p(D)$, $C_p(l)$ is the set of all pairs $(f, t) \in Q_p(D)$ that are connected in Wikipedia by a path with length up to l. We will use path queries and computes the coverage of path queries for a set of pairs of Wikipedia articles.

The grey box at the middle of Fig. 3 shows the navigational paths and path queries that can be generated for *(Paris, Pierre Curie)* and *(Rosario, Lionel Messi)* Wikipedia pairs in the example. Notice both cases are covered by a same path query (`#from / Category:#from / Category:People from #from / #to`) instead of the fact that they have different navigational paths.

Definition 7 (Path Index). *Given a $C_p(l)$, a Path Index (PI) is a bipartite graph $(PQ, C_p(l), I)$, it represents the coverage of path queries for a set of pairs of Wikipedia articles that are related by a DBpedia property p. PQ is an ordered set of path (descendent order by element degree), $I \subseteq PQ \times C_p(l)$ is the set of edges* *relating elements from PQ with elements from $C_p(l)$; $(pq, v) \in I \Leftrightarrow pq \in PQ \wedge v \in C_p(l) \wedge v$ covers pq. The first path query in PQ is the general representation of the property p in Wikipedia.*

Definition 8 (Rating). *Given a Path Index $PI = (PQ, C_p(l), I)$ and a path query $pq \in PQ$ the rating of the pq in the path index is defined by the degree of pq in the path index bipartite graph: $rating(pq, PI) = |\{e \in C_p(l) : (pq, e) \in I\}|$*

Path Index are useful because they enable us to obtain the following information:

- Which is the most general path query?
- Which path queries cover a connected pair of Wikipedia articles?
- How many pairs are covered by a path query?

The Path Index for the examples of *(Rosario, Lionel Messi)* and *(Paris, Pierre Curie)* is shown at the bottom of Fig. 3. The pink ellipses with the path queries are the PQ set, the yellow ellipses with the Wikipedia pairs are the $C_p(l)$ set, and the arrows in between is the set of edges I. Finally, the *rating* for `#from / Category:#from / Category:People from #from / #to` is 2, and the *rating* for `#from / #to` is 1.

The information disparity between DBpedia and Wikipedia is detected by comparing the difference between the set of connected pairs of resources in DBpedia and those corresponding pairs that are connected in Wikipedia.

Definition 9 (Wikipedia Pair Connection). *Two Wikipedia articles (a, b) that are related by a DBpedia property with one-to-many cardinality are connected when at least one of the following conditions is true:*

1. *There is a navigational path from a to b through the category tree with length less or equal to five [5, 7].*
2. *There is a direct link from a to b.*
3. *a has a direct link to a **List of** page that has a direct link to b.*

For instance, `Rosario,_Santa_Fe` and `Lionel_Messi` are connected according to the first condition since there is navigational path:`Rosario,_Santa_Fe / Cat:Rosario,_Santa_Fe / Cat:People_from_Rosario,_Santa_Fe / Lionel_Messi`.

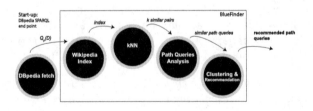

Fig. 4. BlueFinder algorithm steps

Rosario,_Santa_Fe / Lionel_Messi are connected according to the second condition, and, finally, Al_Pacino / List_of_awards_and_nominations_received_by_Al_Pacino / Academy_Award connects the pair elements *(Al_Pacino, Academy_Award)* following the third condition.

4 Problem Statement

We describe the problem of defining the best representation of missing links in Wikipedia as a collaborative recommender system problem. According to Adomavicioius and Tuzhilin [23], *"collaborative recommender systems try to predict the utility of items for a particular user based on the items previously rated by other users"*. BlueFinder predicts links between Wikipedia articles based on links previously rated by the Wikipedia community. Thus, BlueFinder can be considered as a collaborative recommender system for enhancing content of Wikipedia.

More formally, the utility function $u(c, s)$ of item s for user c is estimated based on the utilities $u(c_j, s)$ assigned to item s by those users $c_j \in C$ who are "similar" to user c. In the context of Wikipedia, BlueFinder does not directly apply recommenders to suggest Wikipedia articles to users but to suggest links between articles. BlueFinder predicts the utility of path queries for a particular pair of Wikipedia articles based on those rated by the Wikipedia community. In other words, the pairs of articles *(from, to)* will play the role of users and the path queries will be the items. Then, the utility $u(c, pq)$ of a path query pq for a pair c related by a semantic property p is estimated based on the utilities $u(c_j, pq)$ assigned to pair c by those pairs $c_j \in C_p(l)$, $u : Q_p(D) \times PQ \to R$, where R is a list of path queries sorted according to the rating (see Definition 8).

Given a property p in DBpedia, $C_p(l)$ and PQ path queries covered by the elements of $C_p(l)$. Then, for a given pair of Wikipedia articles $(from, to)$, we have to recommend the path query that maximizes the utility function. The following use case illustrates this problem statement in a practical use.

Use Case 1. A user would like to know *which is the best convention to represent the* is birthplace of *semantic relation for the pair of articles* (Boston, Robin Moore) *in Wikipedia*. The expected result is a list of recommended navigational paths that could connect Boston and Robin Moore articles respecting Wikipedia conventions. Some possible answers could be the following navigational paths:

– Boston / * / Category: Writers from Boston / Robin Moore (High confidence)
– Boston / Robin Moore (Low confidence)

The first specifies a navigational path that starts in Boston article, then it could continue by several other articles but it has to finish in the category *Writers from Boston* and finally *Robin Moore* article. The second navigational path specifies the direct navigation from *Boston* to Robin Moore. Additionally, the first respects the Wikipedia convention with a higher level of confidence than the second one.

Algorithm 1. BlueFinder

Require: x : unconnected pair, $maxR$: maximum number of recommendations, $Q_p(D)$, k : number of neighbors, l:max path length

Ensure: Recommendation path query set

1: $index = (PQ, C_p(l), I) \leftarrow WikipediaIndex(Q_p(D), l)$
2: $k_{neighbors} \leftarrow kNN(x, C_p(l))$
3: $knnPQ \leftarrow \bigcup_{c_i} pq : (pq, c_i) \in I, c_i \in k_{neighbors}$
4: $knnI \leftarrow \bigcup_{c_i} (pq, c_i) : (pq, c_i) \in I, c_i \in k_{neighbors}$
5: $knnPI \leftarrow (knnPQ, k_{neighbors}, knnI)$
6: $M \leftarrow NoiseFilter(knnPI)$ {M ordered in rating descendent order}
7: $M \leftarrow StarGeneralization(M, knnPI)$
8: **return** $maxRecom$ path queries of M

---

5   BlueFinder

BlueFinder implements a four-steps pipeline process as shown in Fig. 4. A pre-processing step *DBpedia fetch* configures the BlueFinder start-up information. It fetches from DBpedia SPARQL endpoint the set of pairs of Wikipedia articles $Q_p(D)$ that are related in DBpedia by a semantic property p. After having the $Q_p(D)$, BlueFinder algorithm is ready to start.

The BlueFinder Algorithm 1 receives five inputs: (1) the unconnected pair of Wikipedia articles x, (2) maximum number of recommendations $maxR$, (3) the $Q_p(D)$ set generated by *DBpedia fetch* step, (4) the number of neighbors k , and (5) the maximum length of a path l. BlueFinder algorithm starts by invoking the *WikipediaIndex*.

The *WikipediaIndex* Algorithm 2 builds a path index. It receives $Q_p(D)$ and computes the item set, user set and item ratings. The items are the path queries, and the users are the pairs of Wikipedia pages retrieved from DBpedia. In this case, for each pair of Wikipedia articles $(from, to)$ included in a given $Q_p(D)$, the algorithm performs a depth-first search up to l starting from the *from* article and finishing in the *to* article in the Wikipedia graph (lines 1–4). For each reaching *to* article, it generalizes a path and builds the path index as a bipartite graph (lines 5–8). Finally, it returns the path index that is ready to be used in the next step of the BlueFinder Algorithm 1. BlueFinder traverses Wikipedia graph starting from the *from* article and finishing in the *to* article in the Wikipedia graph until the maximum length of a path; consequently, a depth-first search is more appropriate than the breadth-first search for building path index.

After indexing, BlueFinder performs the *kNN* step (line 2 in Algorithm 1). In this step, given a disconnected pair of articles in Wikipedia, BlueFinder identifies the k nearest connected pairs to the disconnected one.

The *kNN* algorithm uses a similarity measure function to select the nearest neighbors. We define the *Semantic Pair Similarity Distance (SPSD)* function to measure the similarity between pairs of article. SPSD is based on the well-known Jaccard distance [24], it measures the degree of overlapping in the DBpedia

Algorithm 2. WikipediaIndex

Require: $Q_p(D)$, l: path length
Ensure: PI bipartite graph
1: $index = (\varnothing, \varnothing, \varnothing)$
2: **for all** $(from, to) \in Q_p(D)$ **do**
3:      $allPaths \leftarrow \varnothing$, $curL \leftarrow 0$, $curPath \leftarrow \varnothing$
4:      $GenerateAllPaths(from, to, l, curL, allPaths, curPath)$
5:      **for all** $path \in allPaths$ **do**
6:          $pathQuery \leftarrow BuildPathQuery(path, from, to)$
7:          $index \leftarrow InsertInIndex(index, pathQuery, (from, to))$
8:      **end for**
9: **end for**
10: **return** $index$

Algorithm 3. GenerateAllPaths

Require: $from, to$: Wikipedia article, $l, curL$: integer, $allPaths$: $setOfPaths$, $curPath$: $path$
Ensure: All paths that start in $from$ and end in to in Wikipedia Graph with length up to 1. The results only include paths through the category tree, the use of List_of_ pages or direct links.
if $from = to$ then
    $allPaths \leftarrow allPaths \bigcup \{curPath\}$
else if $l > curL$ then
    {Traverse through Wikipedia graph edges set E}
    for all $neighbor \in \{n : (from, n) \in E\}$ do
        $curPath \leftarrow curPath + neighbor$
        $curL \leftarrow curL + 1$
        $GenerateAllPaths(neighbor, to, l, curL, allPaths, curPath)$
        $curPath \leftarrow curPath - neighbor$
        $curL \leftarrow curL - 1$
    end for
end if
return $allPaths$

---

types that describe a pair of Wikipedia articles. The range of the SPSD is from 0 (identical pairs) to 1 (totally disjoint pairs). The Semantic Pair Similarity Distance function is defined as:

Definition 10 (Semantic Pair Similarity Distance (SPSD)). *Given two pairs of pages* $c_1 = (a_1, b_1)$ *and* $c_2 = (a_2, b_2)$. *Let* t_{a_1}, t_{b_1}, t_{a_2}, t_{b_2} *data types in DBpedia for* a_1, b_1, a_2 *and* b_2 *respectively. Data types are defined as:*

$$t_{a_1} = \{t : < a_1 \; rdf:type \; t > \in DBpedia\}, \; t_{b_1} = \{t : < b_1 \; rdf:type \; t > \in DBpedia\},$$
$$t_{a_2} = \{t : < a_2 \; rdf:type \; t > \in DBpedia\}, \; t_{b_2} = \{t : < b_2 \; rdf:type \; t > \in DBpedia\}.$$
$$SPSD(c_1, c_2) = \frac{J(t_{a_1}, t_{a_2}) + J(t_{b_1}, t_{b_2})}{2}$$

where J *is Jaccard distance between* c_1 *and* c_2. $J(c_1, c_2) = \frac{|c_1 \cup c_2| - |c_1 \cap c_2|}{|c_1 \cup c_2|}$.

To illustrate, we consider two pairs of Wikipedia pages $c1 = (Paris, Pierre\ Curie)$ and $c2 = (Paris, Larusso)$. The data types are:

- $t_{paris} = \{EuropeanCaptialsOfCulture, PopulatedPlace\}$.
- $t_{PierreCurie} = \{Scientist, FrenchCheimists, PeopleFromParis\}$.
- $t_{Larusso} = \{Artist, PeopleFromParis\}$.

$$SPSD(c_1, c_2) = \frac{J(t_{paris}, t_{paris}) + J(t_{PierreCurie}, t_{Larusso})}{2} = (0 + 0.75)/2 = 0.375$$

Now, we can define the kNN [25] in our context as:

Definition 11 (kNN). *Given a pair $r \in Q_p(D)$ and an integer k, the k nearest neighbors of r denoted $KNN(r, Q_p(D))$ is a set of k pairs from $Q_p(D)$ where $\forall o \in KNN(r, Q_p(D))$ and $\forall s \in Q_p(D) - KNN(r, Q_p(D))$ then $SPSD(o, r) \leq SPSD(s, r)$.*

Having the kNN step computed, the *Path Queries Analysis* step starts. It obtains the path queries that connect the k neighbors in a smaller path index than the original (from line 3 to 5 in Algorithm 1). Indeed, having a $PI = (PQ, C_p(l), I)$; the value for an unknown rated $r_{c,s}$ for unconnected pair in Wikipedia c and a path query $s \in C_p(l)$, can be computed as:

$$r_{c,s} = degree(s, PI') + featured(pq, PI')$$

where $PI' = (PQ, C_p(l)', I)$ and $C_p(l)' = KNN(c, C_p(l))$ and $featured(pq, PI') = \beta$ if $degree(pq, PI') = |C_p(l)'|$, otherwise $featured(pq, PI') = 0$. β is a scalar used to promote those path queries that are a convention shared for all the k neighbors[10] and we call *featured predictions*. BlueFinder has a high level of confidence in *featured predictions*.

The generated path index contains the path queries that will be recommended and its ratings. Before the recommendations are returned, in the step *Clustering and Recommendation* in Fig. 4, BlueFinder cleans regular-user-unreachable-paths (e.g., paths that include administrative categories) by means of the noiseFilter (Algorithm 4) and similar path queries are grouped by StarGeneralization algorithm (Algorithm 5). Finally, BlueFinder returns the $maxRecom$ best ranked path queries.

The $NoiseFilter$ Algorithm 4 deletes all the paths queries that are not accessible by a Wikipedia user. Wikipedia includes several administrative categories which are used by administrators. In order to recommend path queries that can be utilized by regular users, $NoiseFilter$ deletes those categories whose names begin with "Articles_", "All_Wikipedia_", etc., such as `Cat:Articles_to_be_merged`.

BlueFinder filters path queries into star path queries in order to reduce data sparsity.

Definition 12. *A star path query $PQ^*(f, t)$ is a group of similar path queries that meet the following rules: (1) $PQ^*(f, t)$ starts with #from and ends with*

---

[10] In this work, we use $\beta = 1000$.

Algorithm 4. NoiseFilter

Require: $PI = (PQ, C, I)$: Path index
Ensure: Set of regular user navigable path queries.
   $noise$ $=$ $\{"Articles\_", "All\_Wikipedia\_", "Wikipedia\_", "Non$ $-$
   $free", "All\_pages\_", "All\_non"\}$
   for all $pq = (p_1, .., p_n) \in PQ;$ **do**
      if p_i contains any c $\in noise; 1 \le i \le n$ **then**
         $PQ \leftarrow PQ - \{pq\}$
      end if
   end for
   return PQ

#to. *(2) The * element can only be placed between **#from** and **#to** variables and * represents any path query. (3) The * cannot be the penultimate element in the path query because it has to make explicit the last part of the path in order to make the connection with the **#to** page.*

Example 1. $PQ^*(f, t)$ =#from/*/Cat:People_from_#from/ #to is a star path query. $PQ^*(f, t)$ =#from/*/#to is not a star path query.

The *StarGeneralization* Algorithm 5 groups path queries into a star path query, if possible.

Solving the Use Case 1. We start by fetching from Wikipedia all the pairs of Wikipedia articles with the form (Place, Person) that are related by the semantic relation *is birthplace of*. The SPARQL query evaluated in the DBpedia endpoint is Q1 (Listing 1.2) and the result is the $Q_p(D)$ set. Later, we initialize the BlueFinder algorithm with the following parameters: (Boston, Robin Moore) as the unconnected pair x, the limit of recommendation with $maxR = 3$, in order to simplify the example only 2 neighbors ($k = 2$), and the maximum length of pairs with 5. In Sect. 6 we evaluate the behavior of BlueFinder with a combination of several values of k and $maxR$.

The Wikipedia Index step generates a similar but larger index than the one exemplified in Fig. 3. This *index* includes 409,812 connected pairs $(C_p(l))$, 65,262 path queries (PQ) and 648,631 edges (I). With the index, the next step selects the 2 connected pairs of articles that are most similar to (Boston, Robin Moore). The SPSD detects (Boston, Ursula Parrot) and (Boston, Tom Kratman) as the most similar pairs because both of them are sharing most of the DBpedia types with (Boston, Robin Moore). For example all of them are writers from Boston. After that, BlueFinder only analyzes the path queries that cover the k neighbors in a smaller index as shown in Fig. 5(a). Those similar path queries are generalized into star path queries in the *Clustering and Recommendation* step. The generalization generates the index shown in Fig. 5(b).

Finally, the path query #from / ... / Category: Writers from #from / #to will be a featured recommendation because it covers all the neighbors.

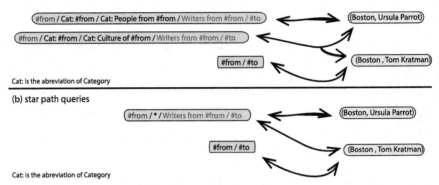

Fig. 5. BlueFinder: similar path queries and star path queries in the execution of use case 1

The other path query covers only one. Then, it retrieves the following recommendations customized for Boston and Robin Moore.

- Boston / ... / Category: Writers from Boston / Robin Moore (Featured Recommendation - High confidence)
- Boston / Robin Moore (Low confidence)

With this recommendations, the user publishes in Wikipedia a navigational path that starts in *Boston* article, and has to finish in the category *Writers from Boston*. For this case, the user only has to add the article *Robin Moore* to the category *Writers from Boston*.

6 Evaluation

In this section we analyze the behavior of our approach by means of measuring the prediction the accuracy of BlueFinder predictions over the 20 properties shown in Table 1. The evaluation is conducted to answer the following questions:

1. What is the best combination of k and $maxRecom$ values to observe the best accuracy from BlueFinder?
2. Does BlueFinder retrieve path queries that can fix missing relations in Wikipedia?
3. Does the confidence level provided by BlueFinder correlate with the accuracy of the predictions?
4. Does the Wikipedia Community use different conventions to represent a DBpedia property?

In this section we describe the method of the evaluation, the evaluation metrics, and then the data sets used in the experimentation are presented. Finally, the results and discussions are introduced.

Algorithm 5. StarGeneralization

Require: PQ: set of path queries, PI: Path index
Ensure: PQ^*: set of star path queries
  $PQ^* \leftarrow \emptyset$
  for all $pq = (p_1, .., p_{n-1}, p_n) \in PQ$; **do**
    if p_{n-1} starts with "`Cat:`" **then**
      $PQ^* \leftarrow PQ^* \bigcup \{(p_1, *, p_{n-1}, p_n)\}$
    else
      $PQ^* \leftarrow PQ^* \bigcup \{pq\}$
    end if
  end for
  return PQ^*

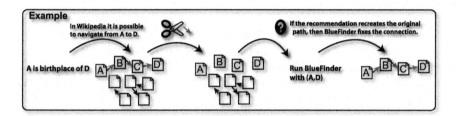

Fig. 6. Evaluation method

6.1 Method

In order to answer the questions described in the previous section, an offline evaluation was designed; user interaction is not considered in the study. The central idea of this evaluation is based on disconnecting connected pairs of articles in Wikipedia and then observing whether BlueFinder is able to recreate them. The important fact here is that BlueFinder has to recreate the Wikipedia community conventions that were defined to connect the pairs and not only to discover the disconnection. This approach is based on the assumption that all connected pairs in Wikipedia follow *Wikipedian* conventions. Figure 6 summarizes the idea of the evaluation method. For the purpose of this evaluation all the path queries that connect a pair of pages that are related in DBpedia by a property p, are considered the correct paths that represent the property p.

A sample of 10 % of the connected pairs was taken for the evaluation. They are randomly selected and kept in a set called N. For instance, for $prop_1$ in Table 1, 188,324 pairs are connected in Wikipedia (i.e. 409,812 - 221,788), so 18,832 randomly selected of those pairs will be in the set N. After that, for each connected pair (w_1, w_2) in N the evaluation repeats the following steps:

1. All paths currently connecting (w_1, w_2) in Wikipedia are stored in the $\mu_{relevant}$ set, and immediately all them are eliminated from Wikipedia to disconnect (w_1, w_2).

2. BlueFinder is executed to predict the paths that could connect (w_1, w_2). The resulting predictions are kept in $\mu_{predicted}$.
3. The $\mu_{predicted}$ set is compared with $\mu_{relevant}$ set in order to compute the metrics detailed below such as precision, recall and F1.
4. Finally, Wikipedia is restored up to the state before the pair disconnection. This means that the (w_1, w_2) pair is reconnected by means of $\mu_{relevant}$.

In this evaluation, BlueFinder behavior is evaluated in each property mentioned in Table 1, and then aggregates the values of all the metrics to have a general point of view. For example, the evaluation measures the precision metric for $prop_1$, then for $prop_2$ and then it continues with the rest of metrics and properties. After all the metrics and properties are computed, the mean of all metric values is calculated.

In order to have an analysis of the best combination of the number of neighbors and the number of the BlueFinder recommendations, the BlueFinder execution is configured with many combinations of the parameter k and $maxRecom$ for each disconnected pair. The values for k are from 1 to 10, and the values for $maxRecom$ are 1, 3, 5 and *unlimited*. The limit of path queries l was fixed in 5 according to the analysis presented previously.

Evaluation Metrics. We measured the accuracy of BlueFinder predictions based on the standard metrics of *Precision* (P), *Recall* (R), *F-measure*(F_1) and *hit-rate*.

Precision relates the number of correct path queries that are predicted by BlueFinder to the total of recommended path queries.

$$P = \frac{|\mu_{relevant} \cap \mu_{predicted}|}{|\mu_{predicted}|} \tag{1}$$

where $\mu_{predicted}$ is the set of predicted path queries and $\mu_{relevant}$ is the set of expected path queries.

Recall computes the ratio of expected path queries to the total of recommended path queries.

$$R = \frac{|\mu_{relevant} \cap \mu_{predicted}|}{|\mu_{relevant}|} \tag{2}$$

F_1 *score* is the combination of precision and recall.

$$F_1 = 2 \times \frac{P \times R}{P + R} \tag{3}$$

We also use the hit-rate recommendation accuracy [26, 27] that measures the number of cases where BlueFinder recommends at least one correct path query.

$$hit - rate = \begin{cases} 1 & if |\mu_{relevant} \cap \mu_{predicted}| > 0 \\ 0 & otherwise \end{cases} \tag{4}$$

The previous measures are extended by studying the distribution of path queries predicted by BlueFinder. In this work, we measure the statistical dispersion of each path query i according to the proportion $p(i)$ of a pair coverage using Gini index.

The *Gini index* [28] measures the distribution of recommended path queries. A value close to 0.0 indicates that all path queries are equally recommended while a value close to 1.0 represents that a particular path query is recommended always.

$$GI = \frac{1}{n-1} \sum_{j=1}^{n} (2j - n - 1)p(i_j) \tag{5}$$

where $i_1, ..., i_n$ is the list of path queries ordered according to increasing $p(i)$.

Finally, we combine the previous metrics to analyze the confidence of the predictions. A high level of confidence in a prediction means that the system trusts its prediction while a low confidence means the opposite [29]. BlueFinder determines confidence in two levels: *featured predictions*, and the position of each prediction in the recommendation set. In order to evaluate the confidence, we will compare the confidence with the hit-rate of each prediction.

Limitations. In statistical terms, the $\mu_{relevant}$ set is used as the gold standard for each pair (w_1, w_2) in N. However, $\mu_{relevant}$ could contain paths that are not related to a property in Table 1, or even a potentially correct prediction could be absent in the $\mu_{relevant}$ set. For example, the $\mu_{relevant}$ set for the pair (London , Richard_Blanshard) with the property `deathPlace` was `#from / * / Cat:People_from_#from / #to` and the first two predictions in the prediction set were `#from / * / Cat:People_from_#from / #to` and `#from / * / Cat:Death_in_#from / #to`. The second prediction could be correct but, as it is not included in $\mu_{relevant}$, the evaluation rejects it as a correct one. Taking into account these considerations, the $\mu_{relevant}$ set is an estimation of the actual path queries and in consequence the BlueFinder is evaluated in the context of the worst case.

Datasets. We evaluate BlueFinder with the twenty semantic properties detailed in Table 1. For each property denoted by $prop_i$, a SPARQL query was evaluated on the DBpedia SPARQL endpoint. The SPARQL query for each property follows the template showed in Listing 1.3 and the values of *DBpediaSemanticProperty*, *fromType* and *toType* are replaced in each property scenario for the specific values of the first, second and third column respectively that are detailed in Table 1. For instance, the SPARQL query in Listing 1.2 corresponds to $prop_1$. The number of the Wikipedia connected pairs of each property is the difference between the numbers of the DBpedia connected pairs minus the number of the Wikipedia disconnected pairs (columns fourth and fifth of Table 1). The evaluation was run with a local copy of the English Wikipedia and DBpedia download in July 2013 and they were stored in a MySQL database.

```
prefix  db-owl:<http://dbpedia.org/ontology/>
prefix  db-p:<http://dbpedia.org/property/>
prefix   foaf:<http://xmlns.com/foaf/0.1/>
select  ?fr    ?to
where   { ?db_from a <fromType>    .
             ?db_to a <toType>    .
             ?db_to db-p:<DBpediaSemanticProperty> ?db_from  .
             ?db_from foaf:isPrimaryTopicOf ?fr  .
             ?db_to    foaf:isPrimaryTopicOf ?to
          }
```

Listing 1.3. SPARQL query template for evaluation scenarios.

Evaluation results are described and discussed in the next section. The complete values of all the metrics values with the different values for k and $maxRecom$ of this evaluation are in https://sites.google.com/site/bfrecommender/publications/.

6.2   Results and Discussion

We start by explaining the information gap presented in Table 1, then we report results for each evaluation metric.

Gap Analysis. The last column in the in the Table 1 presents the gap of missing information in Wikipedia. By analyze the ratio of the gap, shown in Fig. 2, we noticed that 9 out of 20 properties have more than 50 % of missing information. In addition, the number of disconnected pairs of the properties $prop_1$ (birthplace), $prop_2$ (deathplace), and $prop_{13}$ (recordlabel) shown in the Table 1, is equivalent to more than the 50 % of all disconnected pairs of other properties.

The smallest gap was in $prop_{20}$ (notableWork) and $prop_{12}$ (previousWork) with only 5 %. In both cases, this is because the links represent basic information of the connected articles and they are expressed as direct links (#from / #to) between the articles.

Accuracy. To assess the best behavior of BlueFinder, we analyze the values of accuracy metrics for the 20 properties from a general perspective. Figure 7 shows four line-charts with the mean values of $precision, recall, F1$ and $hit-rate$ obtained for each property. Each chart describes the relation between $maxRecom$ and k values for each metric.

BlueFinder is able to find, on average, between 75 % and 82 % of the relevant paths, and according to the hit-rate values it is able to fix around 88 % of the cases for k greater than 4 and $maxRecom = 3$, 5 or $unlimited$. However, the limitations is that the precision values decrease according to the variation of the k values and the number of recommendations.

To detect the best correlation between precision and recall we use the F1 metric. According to the Fig. 7, all the $maxRecom$ curves converge at $k = 5$ with

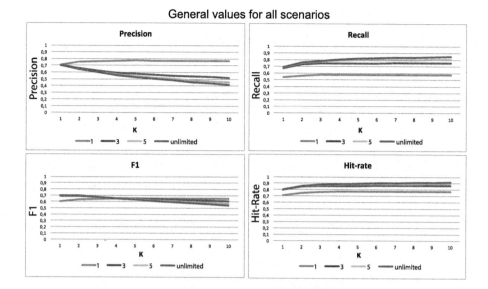

Fig. 7. Precision, Recall, F1, and Hit-rate mean of all properties

value 0.65. Therefore, $maxRecom = 5$ and $k = 5$ determine the best accuracy for BlueFinder. The number of correct path queries tips the scales in favor of recall and hit-rate rather than precision. This assumption is based on the fact that the recommendations are presented to the users in descending confidence order, and consequently, the users have extra information to determine the accuracy of the recommendation. Finally, the unlimited $maxRecom$ was dismissed because it had similar recall than $maxRecom = 5$ but lower precision.

Precision. As presented in Fig. 8 most of the precision curves decrease, due to BlueFinder introduced non-expected path queries in $\mu_{predicted}$ set. This is because of the size of $maxRecom$ has increased but also because the distant neighbors insert noisy paths. Nevertheless, the 70 % of the properties had precision higher than 0.5 at k = 5 and $maxRecom = 5$. This evidences that in general terms the precision of BlueFinder was considerably good taking into account that, as we have mentioned, the predictions are presented in confidence order bringing to the users better information (more details in Sect. 6.2).

Properties $prop_1$: $birthplace$(• line in charts), $prop_2$: $deathplace$ (line with ×), and $prop_{15}$: $country$ (line with △) have low precision less than 0.44 for any $maxRecom$ value. This is because they have a high number of disconnected pairs and shows up that BlueFinder, as many recommender systems, is sensitive to the sparsity of data. On the other hand, property $prop_{12}$: $previousWork$ (line with +) had a high precision: 0.93 with $maxRecom = 1$ and $k = 5$, but it sharply decreased when the $maxRecom$ increased (0.44 with $maxRecom = 3$, 0.31 with $maxRecom = 5$ and 0.23 with $maxRecom = unlimited$). This is because although BlueFinder was able to predict the correct path query for this property, the other path queries that are included in the prediction set are not correct.

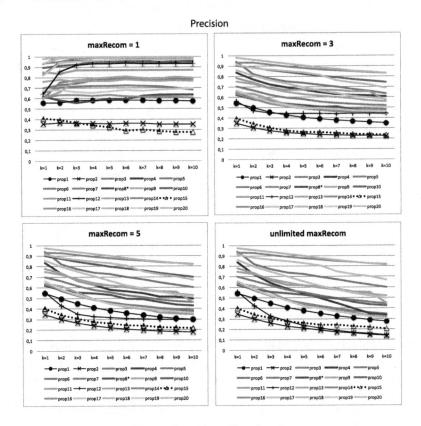

Fig. 8. Precision all properties

Recall and F1. As depicted in Fig. 9, the recall of seventeen properties is greater than 0.7, and the recall of eleven properties is greater than 0.8; all of them with k = 5 and $maxRecom = 5$. Again properties $prop_{12}$, $prop_2$, and $prop_{15}$ are out of norm. The lowest recall value with k = 5 and $maxRecom = 5$ is 0.473 for $prop_2$ property, and the maximum value is 0.972 for $prop_{14}$ property.

Figure 10 presents the valus of F1 metric, although the values for the properties $prop_2$, $prop_{12}$ and $prop_{15}$ are low, twelve properties out of twenty have F1 values greater than 0.6 at k = 5 and $maxRecom = 5$.

Hit-Rate. As depicted in Fig. 11, 80 % of the properties (16 out of 20) have a hit-rate greater than 0.84, and only two properties have values lower than 0.6. The properties with the lowest hit-rate are $prop_{15}$ and $prop_2$; both confirmed the same tendency that appeared in the previous accuracy metric. Although the hit-rate values are low, according to its high level of information gap, the average of hit-rate for the property $prop_8$ is greater than 75 %. These values are promising since 98 % of properties pairs were disconnected in Wikipedia.

Recall

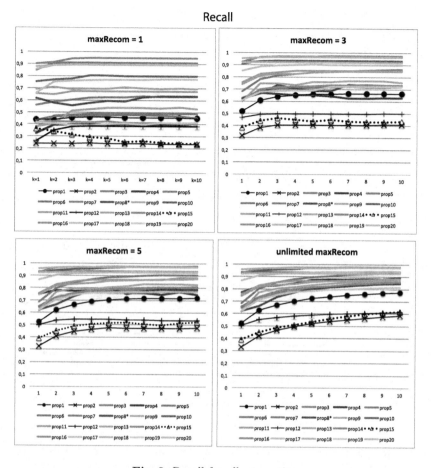

Fig. 9. Recall for all properties

The accuracy values demonstrated that BlueFinder retrieves good recommendations. The hit-rate curves confirmed the best combination of k and $maxRecom$ values by setting $k = 5$ and $maxRecom = 5$.

The BlueFinder predictions are sorted in confidence descendent order; consequently, first ranked predictions may have a better hit-rate than the following ones. In order to answer the third question of the evaluation *Does the confidence level provided by BlueFinder correlate with the accuracy of the predictions?* the hit-rate of BlueFinder featured predictions for the twenty semantic properties is shown in Table 2. As we can see, featured predictions made by BlueFinder are chiefly prominent: the lowest ratio was 0.84 for property $prop_1$ and the highest ratio was 1 for properties $prop_6$ and $prop_{15}$. The mean of all the hit-rate values was nearly 0.95 and the geometric mean was similar ($\simeq 0.95$). This means that BlueFinder is able to fix nearly all the cases where it recommends a featured prediction. Additionally, more than 50 % of the BlueFinder recommendation in this evaluation were featured recommendations. This means, making a projection over the unconnected cases, around 270,367 of new connections in Wikipedia.

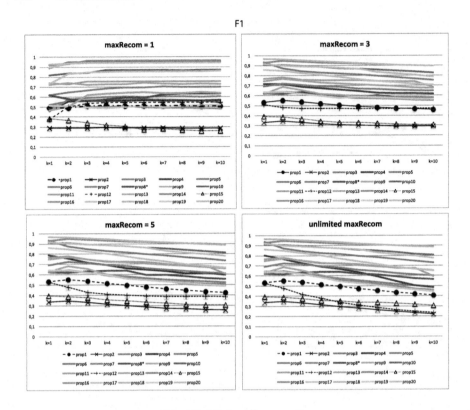

Fig. 10. F1 for all properties

Confidence. Additionally, Fig. 12 extends the information from Table 2 to all the recommendation positions in a line-chart which compares the hit-rate to the first five positions of the BlueFinder recommendation. As we expected, the curve is in descending order while the first position has the best hit-rate (0.78) and the last position the lowest (0.17). This confirms the correlation between the confidence and hit-rate of the predictions. Unfortunately, the hit-rate curve descends more rapidly than we expected to second position and continues descending until the last position.

Distribution of the BlueFinder Recommendations. Wikipedia editors mainly use one convention to represent 18 out of 20 properties in Table 1. BlueFinder is able to predict one path query for these properties; in this case the Gini index is greater than 0.8.

However, because Wikipedia editors use more than one convention for properties such as $prop_9$ and $prop_{11}$, the BlueFinder predictions have Gini index values between 0.474 and 0.778 for different k, and BlueFinder predicts several path queries. For instance, the convention for the property $prop_9 : debutStadium$

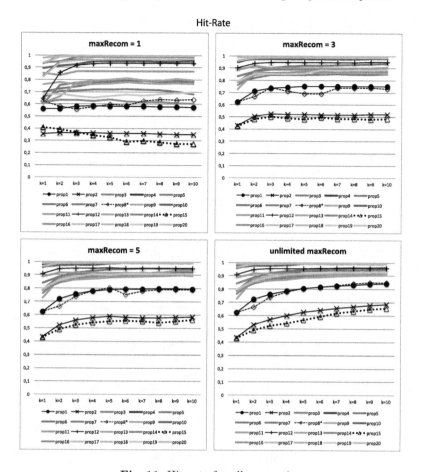

Fig. 11. Hit-rate for all properties

is either a *Standalone list* or a *Category* such that #from / List_of_West_Coast_Players / #to and #from / * / Category: West_Coast_Players / #to.

General Evaluation Conclusions. Evaluation show that the information gap between DBpedia and Wikipedia is a real and important problem. According to the evaluations, the best accuracy of BlueFinder is obtained with $k = 5$ and $maxRecom = 5$, and this answers the first question of the evaluation. With these values, the BlueFinder predictions maximize the expected results with a balanced F1 value.

Additionally, on average 89 % of the disconnected pairs are fixed by BlueFinder according to the hit-rate values and almost all the featured predictions fix the disconnection. A Wikipedia editor could use BlueFinder and fix unconnected pairs in Wikipedia.

Table 2. Confidence and hit-rate for High Confidence predictions

DBpedia Property	Hit-rate
$prop_1$	0.849264
$prop_2$	0.853425
$prop_3$	0.914697
$prop_4$	0.992218
$prop_5$	0.954365
$prop_6$	1
$prop_7$	0.918067
$prop_8$	0.968254
$prop_9$	0.97541
$prop_{10}$	0.977757
$prop_{11}$	0.938776
$prop_{12}$	0.93861
$prop_{13}$	0.950509
$prop_{14}$	0.994295
$prop_{15}$	1
$prop_{16}$	0.868421
$prop_{17}$	0.979371
$prop_{18}$	0.997813
$prop_{19}$	0.938967
$prop_{20}$	0.993056
Mean	**0.95016375**
Geometric Mean	**0.94896**

In order to answer the third question, BlueFinder gives the user the recommendations in descending confidence order. The confidence is also correlated with the hit-rate of the prediction. This enables users to make a better choice of the predictions. The hit-rate of the predictions in the first position is accurate and it is also better when the prediction is a *featured prediction*. Wikipedia editors, in general, only use one convention to represent links in Wikipedia but in some cases; two in this evaluation; some communities define particular conventions. We can conclude this by taking into account that the prediction distribution is centralized in one convention. Additionally, the evaluations showed that those predictions are accurate.

BlueFinder gives good predictions even when the contributors have different conventions. Indeed, this can be concluded by taking into account that the Gini Index, in most of the cases, defined a centralized distribution of a path query in the recommendations, and also the accuracy level of BlueFinder in those cases (Tables 3 and 4).

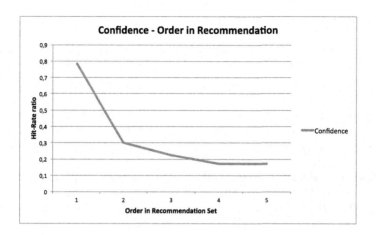

Fig. 12. Confidence and hit-rate according to prediction order in the recommendation set

Table 3. Gini index of the properties

DBpedia Property	K=1	K=2	K=3	K=4	K=5	K=6	K=7	K=8	K=9	K=10
$prop_1$: birthPlace	0.946	0.921	0.902	0.887	0.875	0.866	0.857	0.850	0.843	0.836
$prop_2$: deathPlace	0.922	0.890	0.866	0.849	0.835	0.823	0.814	0.806	0.798	0.792
$prop_3$: party	0.954	0.938	0.928	0.921	0.912	0.906	0.902	0.898	0.894	0.891
$prop_4$: firstAppearance	0.909	0.899	0.885	0.873	0.862	0.854	0.849	0.838	0.835	0.831
$prop_5$: recordLabel	0.956	0.938	0.925	0.912	0.902	0.895	0.889	0.881	0.876	0.871
$prop_6$: associatedBand	0.942	0.929	0.923	0.915	0.907	0.896	0.885	0.872	0.865	0.856
$prop_7$: Company	0.941	0.917	0.901	0.886	0.876	0.872	0.866	0.860	0.855	0.850
$prop_8$: recordedIn	0.861	0.833	0.830	0.813	0.802	0.790	0.769	0.757	0.747	0.736
$prop_9$: debutstadium	**0.676**	**0.648**	**0.613**	**0.594**	**0.595**	**0.581**	**0.576**	**0.561**	**0.545**	**0.524**
$prop_{10}$: producer	0.959	0.944	0.933	0.927	0.921	0.915	0.911	0.908	0.904	0.900
$prop_{11}$: training	**0.778**	**0.729**	**0.641**	**0.629**	**0.626**	**0.594**	**0.585**	**0.578**	**0.570**	**0.474**
$prop_{12}$: previousWork	0.955	0.941	0.938	0.931	0.931	0.931	0.927	0.924	0.921	0.919
$prop_{13}$: recordLabel	0.959	0.948	0.938	0.930	0.925	0.920	0.917	0.912	0.906	0.905
$prop_{14}$: starring	0.988	0.977	0.964	0.953	0.942	0.933	0.926	0.918	0.910	0.904
$prop_{15}$: country	0.943	0.929	0.920	0.908	0.900	0.891	0.886	0.882	0.875	0.873
$prop_{16}$: city	0.960	0.942	0.927	0.914	0.904	0.896	0.889	0.882	0.878	0.874
$prop_{17}$: associatedBand	0.963	0.939	0.926	0.915	0.911	0.904	0.899	0.893	0.889	0.885
$prop_{18}$: fromAlbum	0.967	0.953	0.942	0.932	0.923	0.913	0.901	0.885	0.872	0.869
$prop_{19}$: location	0.967	0.945	0.927	0.911	0.898	0.888	0.881	0.872	0.866	0.861
$prop_{20}$: notableWork	0.976	0.962	0.954	0.950	0.948	0.940	0.933	0.934	0.934	0.931

Table 4. Gini index of the properties $prop_9$: debutstadium and $prop_{11}$: training with $maxRecom=5$

Property	K=1	K=2	K=3	K=4	K=5	K=6	K=7	K=8	K=9	K=10
$prop_9$	0.678	0.655	0.630	0.627	0.643	0.649	0.661	0.652	0.662	0.680
$prop_{11}$	0.778	0.729	0.650	0.636	0.648	0.636	0.623	0.616	0.607	0.509

7 Conclusions and Further Work

In this paper, we introduce the information gap between Wikipedia and DBpedia. To reduce this gap, we have to discover Wikipedia conventions to represent a DBpedia property between a pair of Wikipedia articles. We propose BlueFinder, a collaborative recommender system that recommends navigational paths to represent a DBpedia property in Wikipedia, while respecting Wikipedia conventions. BlueFinder learns from those similar pairs already connected by Wikipedia community and proposes a set of recommendations to connect a pair of disconnected articles. BlueFinder exploits DBpedia types to define a similarity function. Experimental results demonstrate that BlueFinder is able to fix in average 89 % of the disconnected pairs with good accuracy and confidence.

Currently, BlueFinder is tailored for Wikipedia/DBpedia where entities matching are well-defined. However, BlueFinder can be generalized to other datasets with established entities matching.

As a further work, we plan to update Wikipedia with BlueFinder recommendations. We have detected more than 50 % of the recommendation are *featured recommendations*. This means around 270,367 new links will be added to Wikipedia. The future work will be based on a crowdsourcing activity and a monitoring program which evaluates the community agreement of the new connections. Moreover, we are going to adapt this approach in combination with non-English versions of Wikipedia. Finally, we plan to extend the approach to any property in DBpedia in combination with other languages of Wikipedia and to offre the next generation of BlueFinder as a service for any Wikipedia editor.

Acknowledgements. This work is supported by the French National Research agency (ANR) through the KolFlow project (code: ANR-10-CONTINT-025), part of the CONTINT research program.

References

1. Lu, C., Stankovic, M., Laublet, P.: Desperately searching for travel offers? formulate better queries with some help from linked data. In: Gandon, F., Sabou, M., Sack, H., d'Amato, C., Cudré-Mauroux, P., Zimmermann, A. (eds.) ESWC 2015. LNCS, vol. 9088, pp. 621–636. Springer, Heidelberg (2015)
2. Lehmann, J., Isele, R., Jakob, M., Jentzsch, A., Kontokostas, D., Mendes, P.N., Hellmann, S., Morsey, M., van Kleef, P., Auer, S., Bizer, C.: DBpedia - a large-scale, multilingual knowledge base extracted from wikipedia. Semant. Web J. **6**(2), 167–195 (2015)
3. Torres, D., Molli, P., Skaf-Molli, H., Díaz, A.: Improving wikipedia with DBpedia. In: Mille, A., Gandon, F.L., Misselis, J., Rabinovich, M., Staab, S. (eds.) WWW (Companion Volume), pp. 1107–1112. ACM (2012)
4. Pérez, J., Arenas, M., Gutierrez, C.: Semantics and complexity of SPARQL. ACM Trans. Database Syst. **34**(3), 16:1–16:45 (2009)
5. Landauer, T.K., Nachbar, D.: Selection from alphabetic and numeric menu trees using a touch screen: breadth, depth, and width. ACM SIGCHI Bull. **16**(4), 73–78 (1985)

6. Larson, K., Czerwinski, M.: Web page design: implications of memory, structure and scent for information retrieval. In: Proceedings of the SIGCHI Conference on Human Factors in Computing Systems, CHI 1998, pp. 25–32. Press/Addison-Wesley Publishing Co., ACM, New York (1998)
7. Otter, M., Johnson, H.: Lost in hyperspace: metrics and mental models. Interact. Comput. **13**(1), 1–40 (2000)
8. Torres, D., Molli, P., Skaf-Molli, H., Diaz, A.: From DBpedia to wikipedia: filling the gap by discovering wikipedia conventions. In: 2012 IEEE/WIC/ACM International Conference on Web Intelligence, WI 2012 (2012)
9. Torres, D., Skaf-Molli, H., Molli, P., Diaz, A.: BlueFinder: recommending wikipedia links using DBpedia properties. In: ACM Web Science Conference 2013, WebSci 2013, Paris, France, May 2013
10. Wang, Y., Wang, H., Zhu, H., Yu, Y.: Exploit semantic information for category annotation recommendation in wikipedia. In: Kedad, Z., Lammari, N., Métais, E., Meziane, F., Rezgui, Y. (eds.) NLDB 2007. LNCS, vol. 4592, pp. 48–60. Springer, Heidelberg (2007)
11. Mirizzi, R., Di Noia, T., Ragone, A., Ostuni, V.C., Di Sciascio, E.: Movie recommendation with DBpedia. In: IIR, pp. 101–112. Citeseer (2012)
12. Panchenko, A., Adeykin, S., Romanov, A., Romanov, P.: Extraction of semantic relations between concepts with knn algorithms on wikipedia. In: Proceedings of Concept Discovery in Unstructured Data Workshop (CDUD) of International Conference On Formal Concept Analysis, pp. 78–88 (2012)
13. Singer, P., Niebler, T., Strohmaier, M., Hotho, A.: Computing semantic relatedness from human navigational paths: a case study on wikipedia. Int. J. Seman. Web Inf. Syst. (IJSWIS) **9**(4), 41–70 (2013)
14. Di Noia, T., Mirizzi, R., Ostuni, V.C., Romito, D., Zanker, M.: Linked open data to support content-based recommender systems. In: 8th International Conference on Semantic Systems (I-SEMANTICS 2012), ICP, ACM Press (2012)
15. Pereira Nunes, B., Dietze, S., Casanova, M.A., Kawase, R., Fetahu, B., Nejdl, W.: Combining a co-occurrence-based and a semantic measure for entity linking. In: Cimiano, P., Corcho, O., Presutti, V., Hollink, L., Rudolph, S. (eds.) ESWC 2013. LNCS, vol. 7882, pp. 548–562. Springer, Heidelberg (2013)
16. Adafre, S.F., de Rijke, M.: Discovering missing links in wikipedia. In: Proceedings of the 3rd International Workshop on Link Discovery, LinkKDD 2005, pp. 90–97. ACM, New York (2005)
17. Sunercan, O., Birturk, A.: Wikipedia missing link discovery: a comparative study. In: AAAI Spring Symposium: Linked Data Meets Artificial Intelligence, AAAI (2010)
18. Hoffmann, R., Amershi, S., Patel, K., Wu, F., Fogarty, J., Weld, D.S.: Amplifying community content creation with mixed initiative information extraction. In: Proceedings of the 27th International Conference on Human Factors in Computing Systems, CHI 2009, pp. 1849–1858. ACM, New York (2009)
19. Suchanek, F.M., Kasneci, G., Weikum, G.: Yago: a core of semantic knowledge. In: Proceedings of the 16th International Conference on World Wide Web, WWW 2007, pp. 697–706. ACM, New York (2007)
20. Alkhateeb, F., Baget, J.F., Euzenat, J.: Extending SPARQL with regular expression patterns (for querying RDF). Web Seman. Sci. Serv. Agents World Wide Web **7**(2), 57–73 (2011)
21. Abiteboul, S., Vianu, V.: Regular path queries with constraints. In: Proceedings of the Sixteenth ACM SIGACT-SIGMOD-SIGART Symposium on Principles of Database Systems, PODS 1997, pp. 122–133. ACM, New York (1997)

22. Arenas, M., Conca, S., Pérez, J.: Counting beyond a yottabyte, or how SPARQL 1.1 property paths will prevent adoption of the standard. In: Proceedings of the 21st International Conference on World Wide Web, WWW 2012, pp. 629–638. ACM, New York (2012)

23. Adomavicius, G., Tuzhilin, A.: Towards the next generation of recommender systems: a survey of the state-of-the-art and possible extensions. IEEE Trans. Knowl. Data Eng. 17(6), 734–749 (2005)

24. Jaccard, P.: Nouvelles recherches sur la distribution florale. Bull. de la Socièté Vaudense des Sciences Naturelles 44, 223–270 (1908)

25. Lu, W., Shen, Y., Chen, S., Ooi, B.: Efficient processing of k nearest neighbor joins using mapreduce. Proc. VLDB Endowment 5(10), 1016–1027 (2012)

26. Deshpande, M., Karypis, G.: Item-based top-n recommendation algorithms. ACM Trans. Inf. Syst. (TOIS) 22(1), 143–177 (2004)

27. O'Sullivan, D., Smyth, B., Wilson, D.C., Mcdonald, K., Smeaton, A.: Improving the quality of the personalized electronic program guide. User Model. User-Adap. Inter. 14(1), 5–36 (2004)

28. Fleder, D.M., Hosanagar, K.: Recommender systems and their impact on sales diversity. In: Proceedings of the 8th ACM Conference on Electronic Commerce, pp. 192–199. ACM (2007)

29. Shani, G., Gunawardana, A.: Evaluating recommendation systems. In: Ricci, F., Rokach, L., Shapira, B., Kantor, P.B. (eds.) Recommender Systems Handbook, pp. 257–297. Springer, US (2011)

Soft and Adaptive Aggregation of Heterogeneous Graphs with Heterogeneous Attributes

Amine Louati[1]([⊠]), Marie-Aude Aufaure[2], Etienne Cuvelier[3],
and Bruno Pimentel[4]

[1] PSL, Université Paris-Dauphine, LAMSADE CNRS UMR 7243, Paris, France
`amine.louati@lamsade.dauphine.fr`
[2] École Centrale Paris MAS Laboratory, Paris, France
`Marie-Aude.Aufaure@ecp.fr`
[3] ICHEC Brussels School of Management, Brussels, Belgium
`etienne.cuvelier@ichec.be`
[4] Centro de Informatica, Universidade Federal de Pernambuco, Recife, Brazil
`bap@cin.ufpe.br`

Abstract. In the enterprise context, people need to exploit, interpret and mainly visualize different types of interactions between heterogeneous objects. Graph model is an appropriate way to represent those interactions. Nodes represent the individuals or objects and edges represent the relationships between them. However, extracted graphs are in general heterogeneous and large sized which makes it difficult to visualize and to analyze easily. An adaptive aggregation operation is needed to have more understandable graphs in order to allow users discovering underlying information and hidden relationships between objects. Existing graph summarization approaches such as k-SNAP are carried out in homogeneous graphs where nodes are described by the same list of attributes that represent only one community. The aim of this work is to propose a general tool for graph aggregation which addresses both homogeneous and heterogeneous graphs. To do that, we develop a new *soft* and *adaptive* approach to aggregate heterogeneous graphs (i.e., composed of different node attributes and different relationship types) using the definition of Rough Set Theory (RST) combined with Formal Concept Analysis (FCA), the well known K-Medoids and the hierarchical clustering methods. Aggregated graphs are produced according to user-selected node attributes and relationships. To evaluate the quality of the obtained summaries, we propose two quality measures that evaluate respectively the similarity and the separability in groups based on the notion of common neighbor nodes. Experimental results demonstrate that our approach is effective for its ability to produce a high quality solution with relevant interpretations.

Keywords: Graphs · Homogeneous and heterogeneous social networks · Aggregation · Clustering

© Springer International Publishing Switzerland 2016
P. Molli et al. (Eds.): SWCS 2013/2014, LNCS 9507, pp. 145–180, 2016.
DOI: 10.1007/978-3-319-32667-2_7

1   Introduction

Data manipulated in an enterprise context are displayed either structured or unstructured such as e-mails, documents, etc. Graphs are a natural way of representing and modeling such data in a unified manner (structured semi-structured and unstructured ones). The main advantage of such structure resides in its dynamic aspect and its capability to represent relations, even multiple ones, between objects. Users need to visualize different types of interactions between heterogeneous objects (e.g. product and site, customers and products, users interactions like social networks, etc.). In order to analyze these interactions and facilitate their visualization, it is interesting to draw on the graph structure.

However, extracted graphs are often very large, with thousands or even millions of nodes and edges. As a result, it is almost impossible to understand the information encoded in these graphs by mere visual inspection. To facilitate the visualization and data interpretation, it seems interesting to perform an operation of exaggeration. The objective of graph aggregation is to produce small and understandable summaries and highlight communities in the network which greatly facilitates the interpretation. Most existing work [1–6] are rather methods of structural partitioning that completely ignore the attributes associated with nodes making interpretation very difficult. The aggregation should use not only relationships between nodes but also it should take into account any available information that can guide the process and enhance the interpretation according to different perspectives.

In this context, few research works were applied to heterogeneous graphs. In [7], authors proposed a model-based clustering algorithm applied on heterogeneous graphs and addressing two challenges: (1) taking into account objects with incomplete attributes and (2) automatically learn the strengths of different types of links in the graph. The method is based on the following hypothesis: (i) link types are more or less important for a particular purpose, and (ii) from a list of attributes specified by the user, attributes may be partially or not contained in an object, and the value could be missing. Attributes can be numeric of textual. Experiments have been performed on real and synthetic datasets. The choice for determining the best number of clusters k is not addressed in the paper. In [8], authors introduced a graph summarization method namely, k-SNAP that provides useful features to help users to extract and to understand hidden information encoded in large graphs. This method allows them to customize the summaries based on user-selected node attributes and relationships. However, two key components are missing that limit the practical application of this method.

– First, the k-SNAP method deals only with categorical node attributes. However, in real world and especially in a business context, many node attributes are numerical, such as the age of the employees or the number of exchanged e-mails between users in an enterprise network. Simply running the graph summarization method on the numerical attributes will result in summaries with large sizes (at least as large as the number of distinct numerical values). On the other hand, k-SNAP is not practical with the presence of a large number of attributes mainly with multiple modalities because it produces scattered and non informative summaries formed by a large number of groups with small size (at least as large as the cardinal of Cartesian product of all modalities).

- Second, the k-SNAP method is restricted to homogeneous graphs where nodes are characterized by the same list of attributes. Nevertheless, new requirements related to the enterprise context appear. For instance, users need to analyze different types of interactions between heterogeneous objects to exploit them for commercial purposes: sending product recommendations to a targeted customer and guide his preferences, and also for organizational purposes: learn about the roles of individuals (e.g. who hold the information, who is the responsible, who is the expert) and their interactions (e.g. who works with whom). Thus, the retrieved graphs are no longer homogeneous but rather heterogeneous as they are extracted from relational databases [9] containing several kinds of relationships and objects.

Based on k-SNAP method which is a two-step method (*A-compatible grouping* step and (A, R)-*compatible grouping* step), we propose in this work a general tool for graph aggregation which takes into consideration homogeneous as well as heterogeneous graphs. To relax the *A-compatible grouping* step, we propose three approaches applied to heterogeneous graphs. The first approach is based on the definition of Rough Set Theory (RST) using Formal Concept Analysis (FCA). The second approach uses the well known K-medoids clustering method. The third one is based on the hierarchical clustering method. We will explain in Sect. 4 in which case each method is adopted. For (A, R)-*compatible grouping* step, we present two new measures to evaluate the quality of summaries. The first measure is a measure of similarity, and the second one is a measure of separability. Our objective is to realize a more refined evaluation comparing to k-SNAP. We no longer based on the notion of common neighbor groups but rather on the notion of *common neighbor nodes*. To divide a group, we propose a new mechanism based on the notion of *central node*.

The rest of this paper is organized as follows. Section 2 presents a review of the existing graph clustering and aggregation methods. In Sect. 3, we define the main concepts used in this work. In Sect. 4, we describe the *A-compatible grouping* step applied to heterogeneous graphs along with its different approaches. In Sect. 5, we describe the (A, R)-*compatible grouping* step illustrated with a running example that compares k-SNAP quality measure with our two proposed quality measures. Experimental results are discussed in Sects. 6 and 7 concludes and presents perspectives for future work.

2   Related Work

When graphs of extracted social networks are large, effective graph aggregation and visualization methods are helpful for users to understand the underlying information. Graph Aggregation Algorithms produce small and understandable summaries and highlight communities in the network, which greatly facilitates its interpretation.

The automatic detection of communities in a social network can provide this kind of graph aggregation. The community detection is a clustering task, where a *community* is a cluster of nodes in a graph [2,3], such as the nodes of the cluster must be more connected with inside nodes, than with nodes outside of the cluster (more details, see [4,10]).

The first class of clustering algorithms are the *partitional algorithms*, which try to find a partition of a set of data, with a given number of clusters, using jointly, most of the time, similarity or a dissimilarity measures and a quality criterion of the found partition. The most popular partitional algorithm (with several variants), the *k-means clustering* [11], tries to find a partition of the set of objects which minimizes the *sum-of-square* criterion which adds the dissimilarities from each object to the center of its own cluster. Several (di)similarity measures can be defined in the social network context, like those based on the *Jaccard index*, which measures similarity between the sets of *neighbors* of the two nodes, but other measures can be defined [4,10].

Hierarchical clustering algorithms try to organize data into a hierarchical structure, and are divided into *agglomerative* and *divisive* algorithms, depending on whether the partition is coarsened, or refined, at each iteration. The basic idea beyond *agglomerative algorithms* is simple: at the starting point, the objects to cluster are their own classes, and then at each stage we merge the two most similar clusters. Of course a dissimilarity measure between two clusters is mandatory, and for a given dissimilarity measure d between objects, several cluster-dissimilarities exist. The result of the clustering process is a *dendogram*, which can be cut to give one single partition. *Divisive clustering algorithms,* split the data set iteratively or recursively into smaller and smaller clusters, with respect to a quality criterion. The most popular method for divisive hierarchical clustering of social networks uses the notion of edge betweenness [1], because finding the connecting edges between communities is also finding these communities. The algorithm given in [2] splits the network into clusters by removing, step by step, the edge with the higher betweenness value. The use of a stopping criterion which measures the improvement at each step should permit to stop when no improvement is gained with an iteration. In most cases the *modularity* defined by [12] is used. SuperGraph [13] employs hierarchical graph partitioning to visualize large graphs.

Specially designed for graphs, *spectral algorithms* [5] are based on the notion of connected components. These algorithms work with a Laplacian matrix based on the adjacency (or weight) matrix [14,15]. If the graph of the social network contains k completely disjoints communities (*i.e.* without any link between them), called *connected components*, then the k *eigenvectors* associated with the eigenvalue 0 are indicator vectors of the k connected components. If the clusters of the social network do not contain "clean" connected components (*i.e.* if there are links between existing communities), then a simple clustering on the k *eigenvectors* associated with the k least eigenvalues, can retrieve the k communities.

Some other algorithms use statistical methods to study graph characteristics, such as degree distributions [16], hop-plots [17] and clustering coefficients [18]. The results are often useful but difficult to control and especially to exploit. Methods for mining frequent graph patterns [6] are also used to understand the characteristics of large graphs. Washio and Motoda [16] provide an elegant review on this topic.

However, all the previous algorithms use only links between nodes within the graph, and do not take into account their internal values, while classical clustering algorithms applied on tables of values, work only on these values ignoring completely the possible links between individuals. An algorithm which can take into account both kinds of information would be very valuable. Designed for graphical graph aggregation, the k-SNAP algorithm [8] in its divisive version is a two-step algorithm where in the first step, called *A-compatible grouping*, it builds an initial partition based on nodes' attributes. In the second step, called *(A,R)-compatible grouping*, it divides the existing groups according to their neighbors groups until the cardinal of the summary reaches k. The output of the k-SNAP aggregation method is a partition in which groups are homogeneous in terms of attributes and relationships.

All approaches that we presented above are carried out with homogeneous graphs. However, in many cases, graphs can be heterogeneous and formed by different types of objects and consequently, nodes are described by different list of attributes. To overcome this issue, we propose in this work a general tool for graph aggregation (see Fig. 1) which takes into account the heterogeneity of our data. Our tool enables users to freely choose according to the type of graph a scenario of aggregation and produce summaries. The idea is to relax the *A-compatible grouping* step of k-SNAP method based on the definition of Rough Set Theory (RST) using the Formal Concept Analysis (FCA) and K-Medoids and hierarchical clustering techniques applied on the different list of attributes. Moreover, we propose two new quality measures to carry out the (A, R)-compatible grouping step. The first measure is a measure of similarity while the second is a measure of separability. The description of these two steps is shown in the following sections.

3   Concepts Definition

We consider a complex graph model called heterogeneous graph [7,19] which takes into account different types of relationships linking two nodes and where nodes may be characterized by a different list of attributes. As for example, in Fig. 2 there are three different relationships between nodes: member, supervisor, and colleague relationships and three different types of nodes: conference, researcher and student.

Definition 1 (Graph model). *Let $V\{x_1, x_2, ..., x_n\}$ be a set of nodes and let $R = \{R_1, R_2, ..., R_r\}$ be a set of relationships types defined on V, we denote $G =< V, E_1, E_2, ..., E_r >$ a graph where $E_t \subseteq V \times V \, \forall t \in \{1, ..., r\}$ is the set of edges such that $(x_k, x_j) \in E_t$ and $\forall t \neq t', E_t \cap E_{t'} = \emptyset$.*

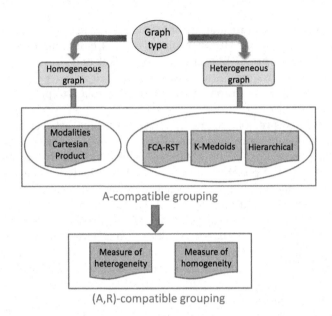

Fig. 1. General tool for graph aggregation

Each element $x_k \in V$ is characterized by a set of $|m_k|$ attributes denoted as $\aleph_{x_k} = \{A_{k_1} = a_{k_1}, ..., A_{k_m} = a_{k_m}\}$, where for each i: $a_{k_i} \in \Re_{k_i}$, i.e. the value a_{k_i} for the attribute A_{k_i} belongs to a given domain $\Re_{k_i}$, and then $\aleph_{x_k} \subset \Re_{k_1} \cup \Re_{k_2} \cup \cdots \cup \Re_{k_m}$. We denote by $\Lambda(G)$ the set of all nodes' attributes in the graph: $\Lambda(G) = \bigcup_{i=0}^{M} \Re_i$, where M is the number of different domain's attributes.

In the graph, the notion of neighborhood of a node can be expressed as follows:

Definition 2 (Node neighborhood). *Let $x \in V$ be a node and $R_t \in R$ be a relationship type, the neighborhood of x regarding the relationship type R_t is defined as: $N_{R_t}(x) = \{y \in V \mid (x, y) \in E_t\} \cup \{x\}$.*

For example, the neighborhood of the node x_9 in Fig. 2 according to colleague relationship is $N_{colleague}(x_9) = \{x_1, x_{10}\}$.

Definition 3 (Partition). *Let $G = (V, E)$ be a graph, $P = \{C_1, C_2, ..., C_p\}$ is a partition on V if and only if all of the following conditions hold:*

- *NonEmpty: $\forall i \in [\![1, k]\!], C_i \neq \emptyset$*
- *Disjoint: $\forall (i, j) \in [\![1, k]\!]^2, i \neq j \Rightarrow C_i \cap C_j \neq \emptyset$*
- *Structure Covering: $\cup_{i=1}^{p} C_i = V$*

For example, a possible partition of V in Fig. 2 is $C_1 = \{x_2, x_7\}$, $C_2 = \{x_4, x_6, x_8, x_{12}\}$ and $C_3 = \{x_1, x_3, x_5, x_9, x_{10}, x_{11}\}$. Each partition represents a different community.

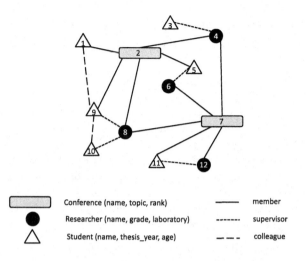

Fig. 2. An example of a heterogeneous graph

We introduce a function denoted Φ_p to control the index assignment of a node in a group. This assignment function is defined as:

Definition 4 (Assignment function). *Let $x_k \in V$ be a node and $P = \{C_1, C_2, ..., C_p\}$ be a partition on V, $\Phi_p : V \mapsto [\![1 \ldots p]\!]$ is an assignment function that assigns an index to x_k indicating the group to which it belongs.*

For example, consider the node x_8 in Fig. 2. According to the previous partition, the index of x_8 is 2.

In this paper, we propose a new mechanism of group division based on the notion of the *central node* of a group.

Definition 5 (Local degree of a node). *Let $x_k \in V$ be a node, $R_t \in R$ be a relationship type and $P = \{C_1, C_2, ..., C_p\}$ be a partition on V, the local degree of a node x_k regarding the relationship type R_t and the partition P is defined as follows:*

$$Deg_{R_t, P}(x_k) = |N_{R_t}(x_k) \cap C_{\Phi_p(x_k)}|$$

It represents the number of neighbors of x_k only within the group $C_{\Phi_p(x_k)}$ according to the relationship R_t.

For example, consider the node x_{10} and *colleague* relationship in Fig. 2. According to the previous partition, $x_{10} \in C_3$. Thus, the local degree of x_{10} is: $Deg_{R_{colleague}, P}(x_{10}) = 1$.

From this definition, we derive a new concept called the complementary local degree including the rest of neighbors of x_i. It is defined as follows:

$$\overline{Deg}_{R_t, P}(x_k) = |N_{R_t}(x_k) \cap \overline{C}_{\Phi_p(x_k)}|$$

In SNA, the greater the degree of a node is, the more central this node is. Consequently, the definition of a central node is as follows:

Definition 6 (Characterization of the *central node* of a group). *Let* $R_t \in R$ *be a relationship type and* $P = \{C_1, C_2, ..., C_p\}$ *be a partition on* V*, a* central node*, denoted* x_α*, of a group* $C_i \in P$ *is defined as follows:*

$$\alpha = argmax_{x_k \in C_i} Deg_{R_t, P}(x_k)$$

In Fig. 2, consider the group $C_3 = \{x_1, x_3, x_5, x_9, x_{10}, x_{11}\}$ of the previous partition and *colleague* relationship. The central node of C_3 is: $x_9 = argmax_{x_k \in C_3} Deg_{R_{colleague}, P}(x_k)$

By introducing the concept of *central node*, we are able to summarize a group via a single node called also prototype, which greatly facilitates the visualization. It should be noted that during the execution of the algorithm (see Algorithm 5), some situations require the introduction of another concept for the centrality to avoid redundancy, named *"the centrality of the second order"*.

Definition 7 (Centrality of second order). *Let* $R_t \in R$ *be a relationship type and* $P = \{C_1, C_2, ..., C_p\}$ *be a partition on* V*, a* central node *of second* order[1]*, denoted* v_β*, is defined as follows:*

$$\beta = argmax_{x_k \in C_i \setminus \{v_\alpha\}} Deg_{R_t, P}(x_k)$$

In Fig. 2, the central node of second order of the group $C_3 = \{x_1, x_3, x_5, x_9, x_{10}, x_{11}\}$ is: $x_1 = argmax_{x_k \in C_i \setminus \{v_9\}} Deg_{R_t, P}(x_k)$

4   *A-compatible grouping* Step in Heterogeneous Graphs

The first and necessary step in k-SNAP is the *A-compatible grouping* step which forms groups of nodes based on the exact matching of their attributes, i.e. after this step, normally, two nodes belonging to the same group have exactly the same values of attributes. In the case of heterogeneous graphs making this kind of grouping can be difficult due to the nature of the data. For example, in a graph (similarly to Fig. 2) where nodes represent researchers, students and conferences, it is expected that researchers have attributes like *name, topics, number of students and name of laboratory* where they work, while students will have attributes like *name, theme of these and name of adviser*. In this case, researchers and students are described by completely different list of attributes, and therefore, it will be impossible to find any exact matching between students and researchers. Considering conferences, we can find attributes like *name, rank and topics*, and then we could find some matching between researchers and conferences based on the shared attribute *topic*, and then not on all attributes. In this work, we start with an approach, based on the basics of FCA and RST, which permits to find *A-compatible grouping* of nodes with heterogeneous attributes.

But this first approach is not completely satisfactory in the case where a large number of groups are found. This is because we do not gain enough aggregation

---

[1] In case of multiple central node of second order, we choose randomly one node.

to perform the *A-R compatible grouping*. Therefore, we propose a second method which permits a *"soft"(less strict) A-compatible grouping using K-Medoids* and allows the user to choose the number k of groups. If this method is a less restrictive technique than the original *A-compatible grouping*, it requires the knowledge of data to fix the parameter k. That is why we propose an original *soft A-compatible grouping using hierarchical clustering* which chooses automatically the number of groups, regarding data.

4.1 A-Compatible Grouping Using RST and FCA

Lets start with an example. In this step, we only deal with attributes of nodes, and not with the links between the nodes, then we are going to focus only on attributes. Suppose that we have a graph with 7 nodes: $\{x_1, x_2, , x_3, x_4, x_5, x_6, x_7\}$ such each node can have, at least, one of the following attributes A_1, A_2 and/or A_3, and suppose finally that these attributes can take, respectively, their values in the following sets: $\{a, b, c\}$, $\{d, e, f\}$ and $\{g, h\}$. Table 1 shows an example of possible values of each attribute for each node. We can remark that not all the nodes have a value for all attributes: nodes x_1 and x_3 does not possess the attribute A_3. The problem is to perform the *A-compatible grouping* step on this kind of data. For this purpose, we use Rough Set Theory (RST) formalism.

RST is a field of set theory whose main study is to analyze inexact, uncertain or vague information [20]. The focus is to create a relationship between rough sets and methods of knowledge discovery, knowledge reduction or acquisition of rules.

Table 1. Table of the attributes values for a graph with heterogeneous attributes.

Attributes	A_1	A_2	A_3
x_1	a	e	
x_2	c	d	g
x_3	a	e	
x_4	c	d	g
x_5	b	d	h
x_6	c	d	g
x_7	b	d	h

A triple (U, A, F) is called an information system if U and A are non-empty sets, respectively of objects and attributes, and F is a set of relations between U and A [21]. The set U is given by $U = \{x_1, \ldots, x_i, \ldots, x_n\}$ where $x_i(i \leq n)$ is an object, and the set A is given as described in the previous section (cf. Definition 1). The last set is given by $F = \{f_j | j \leq M\}$, with $f_j : U \to \Re_j$ and where $\Re_j$ is the domain of the attribute numbered j. And $f_j(x) = a$ means that object x has the attribute numbered j and its value is a.

For a given an information system (U, A, F) and a subset $B \subseteq A$, the binary relation $R_B = \{(x_i, x_j)|f_l(x_i) = f_l(x_j), \forall A_l \in B\}$ is an equivalence relation over U, called the *B-indiscernibility* relation, because if $(x_i, x_j) \in R_B$, then x_i and x_j are indiscernible using the attributes from B. This binary relation is an equivalence relation and then determines a partition $U/R_B = \{[xi]_{R_B}|x_i \in U\}$ where $[x_i]_{R_B} = \{x_j|(x_i, x_j) \in R_B\}$, that is, $[x_i]_{R_B} = \{x_j|f_l(x_i) = f_l(x_j), \forall A_l \in B\}$. So in this framework, doing the *A-compatible grouping* is searching for the partition U/R_B. But the remaining problem is to deal with the heterogeneity of the set of attributes $\Lambda(G)$. In many case we are going to face a comparison between the value of a given attribute for a node against a lack of value for the same attribute for another node. A simple solution is to transform the information system for our nodes with the following binarisation function:

$$F(x, l, a) = \begin{cases} 1 & if \ f_l(x) = a \\ 0 & otherwise \end{cases} \tag{1}$$

using it $\forall x \in V, \forall \Re_l \subset \Lambda(G), \forall a \in \Re_l$.

Table 2. The formal context of the example of Table 1 using the function F given in Eq. (1).

Attributes	A_1			A_2			A_3	
Values	a	b	c	d	e	f	g	h
x_1	1	0	0	0	1	0	0	1
x_2	0	0	1	1	0	0	1	0
x_3	1	0	0	0	1	0	0	1
x_4	0	0	1	1	0	0	1	0
x_5	0	1	0	1	0	0	0	1
x_6	0	0	1	1	0	0	1	0
x_7	0	1	0	1	0	0	0	1

Then we can generate a binary table like the one shown in Table 2, corresponding to the attributes of Table 1. Such a table is used to describe a formal concept and is used in Formal Concept Analysis (FCA).

FCA is a mathematical theory for concepts and concept hierarchies [22], which uses a hierarchical structure called lattice structure and can summarize data into concepts. Consequently, FCA and RST are two different theories which have similar research aim: summarize data. In FCA, a formal context (U, A, I) consists of two sets U and A, and a *binary* relation I $(I \subseteq U \times A)$ between U and A. The elements in U are called the objects and the elements in A are called attributes of the context. In a formal context (U, A, I), if $(x, a) \in I$ it means that "x has a attribute a", or simpler xIa. So it is possible to describe the formal context with a table of 0 and 1, such that, if $(x, a) \in I$, then the value of cell corresponding to the object x (line x) for the attribute a (column a) is 1, otherwise, the value is 0.

Therefore, with the function F given in Eq. (1), the information system is mapped into a formal context and we can define a partition as:

$$U/R_B = \{C_k = \{x_{k_1}, \ldots, x_{k_r}\} | F(x_{k_i}, l, a) = F(x_{k_j}, l, a), \forall x_{k_i}, x_{k_j}, \forall l, \forall a\} \quad (2)$$

in other words we have an A-compatible grouping if the nodes belong to the same equivalence class defined by the function F.

Finally, in the information system framework, a group C_k is such that, whatever the pair of elements x_{k_i} and x_{k_j} that belong to C_k, the list of attributes that describe these two elements is the same and they have the same values, while in the corresponding formal context the list of attributes that describe these two elements is the same and their values belong to $\{0, 1\}$. Note that if an information system framework can be expressed in a formal context using the function (1), conversely a formal context can be seen as an information system where all the attributes as only two possible values : $\{0, 1\}$.

According to Table 2, $[x_1] = [x_3] \Rightarrow C_1 = \{x_1, x_3\}$, $[x_2] = [x_4] = [x_5] \Rightarrow C_2 = \{x_2, x_4, x_5\}$ and $[x_6] = [x_7] \Rightarrow C_3 = \{x_6, x_7\}$. Thus, the partition U/R_A defined by the binary relation R_A on U is given as $U/R_A = \{C_1, C_2, C_3\}$.

Therefore, the representation of information made in FCA and RST can help us to identify groups of nodes represented by different type of attributes where the lists of attributes do not have necessarily the same size. An important procedure of the algorithm that uses this approach is to verify if two nodes are similar according to a set of attributes B. This procedure is described by the Algorithm 1.

This procedure examines if a list of attributes is the same in both objects. The time complexity of this algorithm in the worst case is $O(|B|)$, when it is necessary to verify all attributes in B. Algorithm 2 finds A-compatible grouping in a heterogeneous graph using the FCA-RST approach.

Algorithm 1. $Check(x_i, x_k, B)$

Input: x_i and x_j two objects; B set of attributes.
Result: $isEqual$ a boolean.

1 $isEqual \leftarrow true$;
2 $l \leftarrow 1$;
3 **while** $(l \leq |B|$ and $isEqual == true)$ **do**
4   **if** $(f_l(x_i) \neq f_l(x_k))$ **then**
5       $isEqual \leftarrow false$;
6    $l \leftarrow l + 1$;

7 **return** $isEqual$;

Algorithm 2 starts by generating a group with one node. Then, for every node $x_k \in V$ it verifies if there is a group in which nodes have the same list of

---

Algorithm 2. A-compatible-FCA-RST(G,B)

Input: a graph $G = (V, R)$ where $V = \{x_1, x_2, ..., x_n\}$ is a set of nodes and $R = \{R_1, R_2, ..., R_r\}$ is a set of relationship and a set of attributes $B \subseteq A$

Result: A-compatible grouping ϕ.

1   $C_1 \leftarrow \{x_1\}$;
2   $\phi \leftarrow \{C_1\}$;
3   **for all** $x_i \in V$ **do**
4   **for all** $C_k \in \phi$ **do**
5          $x_k \leftarrow$ first element of C_k;
6          $isEqual \leftarrow Check(x_i, x_k, B)$;
7   **if** *(isEqual == true)* **then**
8             $C_k \leftarrow C_k \cup \{x_i\}$;

9   **else**
10            $C_{|\phi|+1} \leftarrow \{x_i\}$;
11            $\phi \leftarrow \phi \cup C_{|\phi|+1}$;

12 **return** ϕ;

---

attributes of x_k. In case that this group exists, x_k will be attributed to this group. Otherwise, a new group containing this x_k is created. The complexity time to check all nodes is $O(|V|)$ and the one to verify for all groups of ϕ is $O(|V|)$ too, when the list of attributes of some nodes is different. For each node and group, the operation *Check* is executed. Thus, in the worst case, the complexity time of the algorithm is $O(|V|^2|B|)$.

4.2 Soft A-Compatible Grouping Using K-Medoids

The previous approach is useful when nodes in a graph are described by a "grouped list of attributes", i.e. when the formal context table build with the function (1) F is like Table 3. In this last case, two clear groups are formed: $C_1 = \{x_1, x_2, x_3\}$ and $C_2 = \{x_4, x_5, x_6\}$.

However, graphs used to describe real situations can lead to more complex tables. Table 4 highlights this issue. In such cases the final partition would be formed by a large number of small groups. In our example, the result is: $C_1 = \{x_1\}$, $C_2 = \{x_2\}$, $C_3 = \{x_3\}$, $C_4 = \{x_4\}$, $C_5 = \{x_5\}$ and $C_6 = \{x_6\}$. This result is completely unsatisfactory and even if the partition is the result of a clean *A-compatible grouping* step, nodes are not really grouped after this step.

Nevertheless, although nodes have different list of attributes, Table 4 shows a partition composed of two groups $C_1 = \{x_1, x_2, x_3\}$ and $C_2 = \{x_4, x_5, x_6\}$, not based on a strict equality of attributes' values, but based on some similarity between nodes' attributes. The classical way to find such grouping is to use a similarity measure and a clustering method. This is what we propose for graphs of type 2 to find a satisfactory *A-compatible grouping*.

Table 3. Type 1: grouped

	a	b	c	d	e	f
x_1	x	x	x			
x_2	x	x	x			
x_3	x	x	x			
x_4				x	x	x
x_5				x	x	x
x_6				x	x	x

Table 4. Type 2: grouped with different lists

	a	b	c	d	e	f
x_1	x		x			
x_2	x	x	x			
x_3		x	x			
x_4				x	x	
x_5				x	x	x
x_6				x		x

In the literature, many application domains such as pattern recognition, machine learning, data mining, computer vision and computational biology have used clustering algorithms [23–25]. Clustering is an unsupervised learning method whose objective is to group a set of elements into clusters such that elements within a group have a higher degree of similarity, while elements in different groups have a higher degree of dissimilarity. In a practical way, the degree of similarity (or dissimilarity) can be measured using a distance, an angle, a curvature, a symmetry, a connectivity or an intensity with information from the data set [26].

One of the most popular clustering technique is the K-Medoids proposed by Kaufman and Rousseeuw [27], where homogeneous groups are identified by minimizing the clustering error defined as the sum of the squared Euclidean distances between each data set point and the corresponding group prototype. This method is more robust to noise and outliers compared to K-Means and can detect the most important nodes of a given group. However, prototypes in K-Means are means prototypes and most of the time are not nodes of V. A very common realization of K-Medoids clustering is the Partitioning Around Medoids (PAM). Initially, the algorithm defines a partition based on a randomized selection of objects to be the initial prototypes. After, the algorithm repeats two steps: (i) definition of prototypes which correspond to the most representative object (medoid) of each group based on the pairwise dissimilarities; (ii) the allocation step in which objects are swapped to a group such that the distance between an object and the associated prototype minimizes the criterion.

This process finishes when the medoids remain the same or the difference between the actual partition criterion value and the one of the last partition is less than a threshold.

In this work, the adopted similarity coefficient is based on the Jaccard index between the two lists of attributes of the considered objects. To do that, we determine the list of attributes of an object x_j using the operation $x_j{}^* = \{A_l \in B | \exists a \in \Lambda(G) : f_l(x_j) = a\}$, where $B \subseteq \Lambda(G)$ is a set of user-selected attributes. Using the contingency table (see Table 5), it is possible to calculate the following values for two objects x_i and x_j:

- the number of common attributes: $a = |x_i^* \cap x_j^*|$,
- the number of attributes proper to node i: $b = |x_i^*| - a$,
- the number of attributes proper to node j: $c = |x_j^*| - a$,
- the number of lacking attributes for both nodes: $d = |B - (x_i^* \cup x_j^*)|$.

Figure 3 shows a graphic representation of each value a, b, c and d according the sets x_i^*, x_j^* and B.

The chosen similarity coefficient is given by:

$$sim(x_i, x_j) = \frac{a}{a+b+c}. \tag{3}$$

Note that the corresponding dissimilarity is given by:

$$d^t(x_i, x_j) = 1 - sim(x_i, x_j) = \frac{b+c}{a+b+c}. \tag{4}$$

Table 5. Contingency table for shared attributes

	x_j^*	
x_i^*	a	b
	c	d

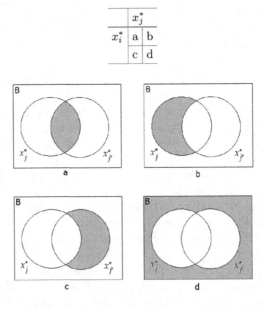

Fig. 3. Graphic representation of a, b, c and d values

Algorithm 3. A-compatible-PAM(G,B,k)

Input: a graph $G = (V, R)$ where $V = \{v_1, v_2, ..., v_n\}$ is a set of nodes and
$R = \{R_1, R_2, ..., R_r\}$ is a set of relationship; a set of attributes $B \subseteq A$
and a number of groups k.
Result: A-compatible grouping ϕ.

1 Let $\epsilon \in \mathcal{R}$;
2 Choose randomly k objects in V to be the initial prototypes;
3 Create k groups and affect each object x_j from V to the group C_i^* whose
  similarity between the object and the prototype is maximized, that is,
  $i^* = argmax_{1 \leq l \leq k} sim(x_j, y_i)$;
4 $W \leftarrow 0$;
5 $W' \leftarrow \sum_{i=1}^k \sum_{x_j \in C_i} sim(x_j, y_i)$
6 **while** $W' - W > \epsilon$ **do**
7   **for all** $C_i \in \phi$ **do**
8         $max \leftarrow \sum_{x'_j \in C_i} sim(y_i, x_{j'})$;
9   **for all** $x_j \in C_i$ **do**
10             $sum \leftarrow \sum_{x'_j \in C_i} sim(x_j, x_{j'})$;
11   **if** $(sum > max)$ **then**
12                 $max \leftarrow sum$;
13                 $y_i = x_j$;
14   Affect each object x_j from V to the group C_{i*} such that
   $i^* = argmax_{1 \leq l \leq k} sim(x_j, y_l)$;
15     $W \leftarrow W'$;
16     $W' \leftarrow \sum_{i=1}^k \sum_{x_j \in C_i} sim(x_j, y_i)$

17 **return** ϕ;

The algorithm used for finding *A-compatible groups* is based on the PAM
algorithm and uses similarity given in Eq. 3. The goal is to maximize the adequacy criterion:

$$W(P, L) = \sum_{i=1}^k \sum_{x_j \in C_i} sim(x_j, y_i) \tag{5}$$

where $sim(x_j, y_i)$ is the similarity between the object x_j and the prototype y_i.
Algorithm 3 shows how to find the A-compatible grouping $\phi = \{C_1, \ldots, C_k\}$.
Let $V = \{x_1, \ldots, x_n\}$ a set of objects and $L = \{y_1, \ldots, y_k\}$ the set of prototypes.
Initially, the algorithm randomly chooses nodes from the set of objects V to be
the initial medoids. Then, each object in V is affected to a group whose similarity to the respective medoid is maximized. After that, the algorithm alternates
iteratively two main steps: definition of medoids and swapping between objects
and prototypes. To find a medoid, a comparison with each object is made: if the
sum of similarities between the chosen object and other objects in the group is
greater than the sum of similarities between this prototype and object of this

group, then this object should be the new medoid, otherwise the actual medoid is unchanged. After the definition of the medoids, the algorithm affects each object to the group which maximizes its similarity with the associated proto-type. Operations of medoids definition and objects affectation are made until the criterion of convergence is satisfied. Traditionally, there are two possibili-ties to stop the algorithm. The first one is to verify if the obtained partition remained unchanged for two successive iterations. Another option is to check if the criterion W reached a stationary value.

The initial partition is defined randomly. For each iteration of the algorithm, the criterion value is archived. The final partition is associated to the best crite-rion. Due to the random step, a partition can be different to another one thus, the result of the execution of this method is not deterministic. In the initial step, it is necessary to make $|V|$ operations to read all objects and $|B|$ operations to calculate the similarity for a pair of object, thus in this step the complexity time is $O(|V||B|)$. After, a comparison between the sum of similarities for each object and the medoids is done. The similarity computation takes $|B|$ operations. As it is necessary do it for $|V|$ objects and k groups, in the worst case a group can have as many objects as the number $|V|$. Thus, the swapping step between objects and prototypes has complexity $O(|V|^2|B|)$. The affectation step is similar to initial step and the complexity is $O(|V||B|)$ too. Finally, the complexity computation of the criterion takes $O(|V||B|)$ operations. Therefore, the algorithm complexity is $O(|V|^2|B|)$.

4.3 Automatic Soft A-Compatible Grouping Using Hierarchical Clustering

In the previous approach, a user has to fix the value of the parameter k to exe-cute the k-medoids approach. Thus, the user should have an *a priori* knowledge about the data in the graph. However, when no knowledge concerning nodes and attributes is known, the hierarchical clustering can be used, since the user does not need the parameter k to execute this kind of method. Therefore, the pro-posal of this approach is to find an *automatic soft A-compatible grouping* using hierarchical clustering.

In hierarchical clustering algorithms, data are not partitioned into one single number of groups, but in a set of hierarchically nested partitions [23], like in a taxonomy, for example,where an object may successively belong to a species, genus, family, order, etc.

The hierarchical clustering techniques family can be divided into two cate-gories: agglomerative algorithms and divisive ones. The agglomerative takes as input the set of n groups (one individual by group) and then merge the groups until data are reduced to a single group containing all nodes, while the divisive techniques make the converse. With the latter, it is necessary to decide at what stage of the algorithm will stop. The advantage of the agglomerative approach is that it is not necessary to define an adequacy criterion to stop the method rather than divisive approach. Therefore, the hierarchical method used in this approach is agglomerative. The similarity measure used for a pair of objects is the one described in the previous subsection (see Eq. (3)).

The hierarchical clustering can be represented by a two-dimensional diagram known as dendrogram (see Fig. 4), which illustrates the coupling or division performed in successive stages of analysis. Each coupling between two objects or two groups of objects is drawn as an horizontal line, and the height of this line gives an information about the similarity of the merged objects: an higher line for a merging means an higher dissimilarity between merged groups. In Fig. 4 we can also see the relationship between the two types of hierarchical algorithms.

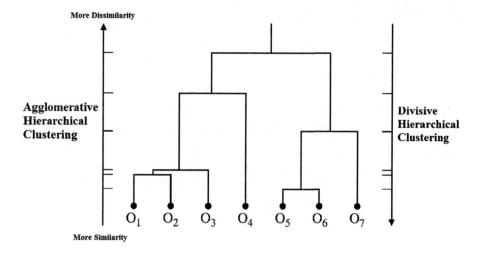

Fig. 4. Agglomerative and divisive hierarchical clustering

An important choice has to be made before performing a hierarchical clustering is to determine how to compute the similarity (or dissimilarity) between two groups which is called the linkage. Table 6 shows the most known types of linkage according to a measure of similarity s (names are referring to the linkage based on a dissimilarity measure):

Table 6. Types of linkages

Name	Expression
Maximum (or complete linkage)	$s(A,B) = min\{s(a,b)\|a \in A, b \in B\}$
Minimum (or single linkage)	$s(A,B) = max\{s(a,b)\|a \in A, b \in B\}$
Mean (or average linkage)	$s(A,B) = \frac{1}{|A||B|} \sum_{a \in A} \sum_{b \in B} s(a,b)$
Centroid	$s(A,B) = s(c_A, c_B), c_A, C_B$ are centroids

Figure 5 shows the sequence of groups junctions according to the agglomerative hierarchical algorithm using average linkage. The first table is the similarity matrix between pairs of nodes $V = \{A, B, C, D, E\}$. It represents also the similarities between the groups, since the agglomerative hierarchical method starts with singleton groups. The equivalent dendrogram is shown in Table 6. An interesting feature of hierarchical clustering resulting in a dendrogram is the fact that we can use the information embedded in this figure to make the final clustering in one partition, and this in two manners: we can choose an height regarding the dendrogram and "cutting it" at this height to create the clusters, or chose the number of clusters and find the corresponding height to find this exact number of groups. Figure 6 shows this: in the height v_1 of the tree, the partition has two groups: $C_1 = \{A\}$ and $C_2 = \{B, C, D, E\}$, whereas, for the value v_2, a new partition of three groups is found: $C_1 = \{A\}$, $C_2 = \{B, C\}$ and $C_3 = \{D, E\}$.

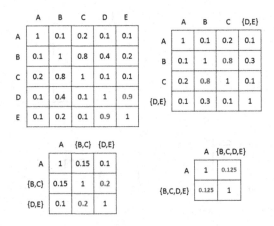

Fig. 5. Matrices of similarities according to average linkage (Color figure online).

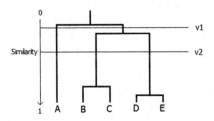

Fig. 6. Cut-similarities v_1 and v_2

It is also possible to automatically find the "right number" of clusters by searching for the knee of a curve. The curve is based on the values of similarities of successive sub-groups merged in the dendrogram (blue values in Fig. 5), and the goal is to find the knee of the curve to find the "optimal number of groups" (see Fig. 7): after the knee the changes are smaller then before it, and then the merging operations are less meaningful. Many methods to find the "ideal"

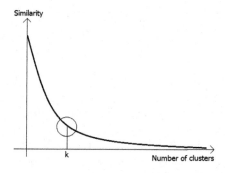

Fig. 7. Curve of similarity versus number of groups

number of groups from a knee of a curve like: the largest magnitude difference between two points, the largest ratio difference between two points, the first data point with a second derivative above some threshold value or data point with the largest second derivative [28]. In this approach, we used the point with the largest second derivative as number of groups. Initially, the similarity of all groups in the agglomerative hierarchical approach is exactly 1 as they have only one object. When a pair of groups is merged, the similarity of this new groups is less than the sub-groups. This union process stops when the method finds one group containing all objects. Thus, the agglomeration process of the algorithm may be read from right to left in the graph.

Algorithm 4 shows the sequence of operations to find the A-compatible grouping using the hierarchical approach. The parameters used in this algorithm are: the graph formed by a set of objects and a set of types of relationships where each node has a list of attributes; a list of user-selected attributes; and a value to cut the tree. This value is not mandatory if the method uses second derivative to find the number of groups. However, it can be used as a flag to indicate if the algorithm will use the second derivative or not. Many strategies can be used to indicate it. In this version of the algorithm, the strategy is: if a value $\in [0,1]$ is given, the algorithm interprets it as a similarity coefficient value and finds the partition ϕ by cutting the tree in the specific height. However, if a value $\notin [0,1]$ is given, the algorithm tries to find the number of groups by second derivative according to the groups' similarities value in the dendrogram.

The algorithm first initializes the list of similarities values that can be used to find the number of groups and the matrix of similarities M. This initial dimension of the matrix is $n \times n$, where n is the cardinality of the set V. These similarities can be calculated as described in the previous section. In each iteration, the algorithm select the two groups that have the largest value of similarity using the matrix M. A new group is created where its sub-groups are the groups selected in the previous step, are removed from list of groups. The largest value is added in the list L. After a new group, the matrix of similarities is updated according to a type of linkage (see Table 6). The update process steps are similar to those shown in Fig. 5. These steps are executed until the number of groups is equal to 1. Finally, the algorithm verifies if the number of groups is reached using the value of similarity *sim* or if it should use the second derivative. After the number of cluster is selected, the A-compatible groups are found, and the grouping is updated.

The time complexity to initialize the matrix M is $O(|V|^2|B|)$, since the calculation of similarity for each pair of objects takes $|B|$ operations and there are $|V|(|V|-1)/2$ combinations of objects. The time complexity to select the pair of groups with largest similarity is $O(|V|^2)$. To add, create and remove groups or update the list L the time complexity is $O(1)$. Whereas, the step to update the matrix M takes $|V|(|V|-1)/2 \times |B| \times |V|$ operations, since it is necessary to compute the similarities, in the worst case, for all $|V|(|V|-1)/2$ pairs of objects using $|B|$ operations and for groups of size $|V|$, thus its time complexity is $O(|V|^3|B|)$. This step can have the complexity $O(|V|^3)$ if an auxiliary matrix of similarity is calculated initially. To compute the second derivative from the list L, it is necessary $|V|$ operations, or time complexity $O(|V|)$. Finally, the time complexity to find the partition using the similarity value is $O(|V|)$. Therefore, the final time complexity in the worst case of this agglomerative hierarchical clustering algorithm is $O(|V|^3|B|)$.

5   (A, R)-*compatible grouping* Step

In this section, we present our tow quality measures, the measure of similarity and the measure of separability. The measure of similarity evaluates locally each group to find the least homogeneous to be divided. The measure of separability evaluates locally the density of each group to find the group with the minimum density to be divided. Prior presenting our quality measures, let us look at how the quality measure k-SNAP is formulated.

5.1   K-SNAP Quality Measure

The quality measure of k-SNAP, denoted Δ, is based on the notion of common neighbor groups. It evaluates the participation ratio of each group relationship (C_i, C_j) for each relationship type R_t [8]. At each iteration, k-SNAP method selects the group that makes the most contribution to the Δ value with one of its neighbor groups according to a particular relationship R_t. This selected group is divided into two subgroups. This mechanism is repeated until the size of the partition is equal to k. To describe the quality measure of k-SNAP, we define the incidence matrix $A_t = (a_{k,l}^t)_{1<k,l<|V|}$ associated with a relationship R_t where $a_{k,l}^t = 1$ iff x_k and x_l are directly connected with an edge of relationship R_t, 0 otherwise. We define also N_{R_t} the participation matrix of rank $|P|(|P|$ is the cardinal of the partition of V) corresponding to a relationship R_t by:

$$(n_{i,j}^t)_{1<i,j<|P|} = \sum_{k=0}^{|C_i|}(1 - \prod_{l=0}^{|C_j|}(1 - a_{kl}^t)) \qquad (6)$$

Algorithm 4. A-compatible-Hierarchical (G,B,sim)

Input: a graph $G = (V, R)$ where $V = \{x_1, x_2, ..., x_n\}$ is a set of nodes and $R = \{R_1, R_2, ..., R_r\}$ is a set of relationship; a set of attributes $B \subseteq A$; sim is the similarity to cut the tree and to find the partition; if $sim \notin [0, 1]$, then the final partition is found according the second derivative.

Result: A-compatible grouping ϕ.

1 Initialize the list $L = \emptyset$, the matrix M of similarities using the coefficients described in the previous section;

2 $numGroups \leftarrow n$;

3 **while** $numGroups > 1$ **do**

4   Select two groups A_{i*} and A_{j*} such that $(i^*, j^*) = argmax_{1 \leq l \leq i \leq numGroups, j < i} M(i, j)$;

5   Create new group $B = A_{i*} \cup A_{j*}$ and add it to D;

6   Remove A_{i*} and A_{i*} from D;

7      $L(numGroups) \leftarrow M(i^*, j^*)$;

8   Update the matrix M using the linkage operation;

9      $numGroups \leftarrow numGroups - 1$;

10 **if** $(sim \in [0, 1])$ **then**

11   Cut the tree of the dendrogram D in the height sim and find K groups. Update the grouping $\phi = \{C_1, ...C_i, ..., C_k\}$;

12 **else**

13   Create a list L' that is the sequence of values corresponding to the second derivative applied to L;

14   Select value K corresponding to the largest value from L';

15   Cut the tree in the height sim0 such that it intersects the tree in K points. Update the grouping $\phi = \{C_1, ...C_i, ..., C_k\}$

16 **return** ϕ;

Then, we define M the matrix of rank $|P|$ which contains the ratios of participation of different groups with respect to the relationship R_t:

$$(m^t_{i,j})_{1<i,j<|P|} = \frac{n^t_{ij} + n^t_{ji}}{|C_i| + |C_j|} \tag{7}$$

For a given graph G, a set of attributes A and a set of relationship R, the quality measure Δ of a the partition P is defined as follows:

$$\Delta(P) = \sum_{1 \leq i,j \leq |P|} \sum_{R_t \in R} \delta^t_{ij} \tag{8}$$

with

$$\delta^t_{ij} = \begin{cases} n^t_{ij} & if \ m^t_{ij} \leq 0.5 \\ |C_i| - n^t_{ij} & otherwise \end{cases} \tag{9}$$

This measure is based on determining the difference in participation of each pair of groups with respect to the relationship R_t. This means that Δ-measure counts the minimum number of differences in participation of group relationships between the given *A-compatible grouping* and a hypothetical (A, R)-*compatible grouping* of the same size. According to Eq. 9 we have two possible cases:

- If the group relationship is weak ($m_{i,j}^t \leq 0.5$), then it counts the extra participants between this weak relationship and a non-relationship ($m_{i,j}^t = 0$).
- Otherwise, *i.e* the group relationship is strong ($m_{i,j}^t > 0.5$), it counts the missing participants between this strong relationship and a 100 % participation-ratio group relationship ($m_{i,j}^t = 1$).

The aim of k-SNAP method is to produce an (A, R)-*compatible grouping* of cardinal k that minimizes the Δ value (initially set to zero). Starting from an *A-compatible grouping*, k-SNAP method selects at each iteration one group to divide. To do that, it computes first for each group C_i its maximal contribution with one of its neighbor groups to Δ. This contribution, denoted CT, is defined as follows: $CT(C_i) = max \{ \delta_{ij}^t \}$. Then, it selects the group having the maximum CT value to be divided into two sub-groups according to the following strategy: the first group contains nodes that have at least a relationship with one node in the group $C_e = \arg \max_{C_j} \{ \delta_{ij}^t \}$ and the other group contains the rest, *i.e.* nodes that do not have relationship with a node in the group C_e.

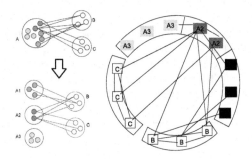

Fig. 8. Graph aggregation using k-SNAP method.

To better understand the principle of k-SNAP method, we run the algorithm on the example of Fig. 8. In this example, nodes have one attribute with three modalities A, B and C. Applying the *A-compatible grouping* step, we obtain a partition composed of three groups in which nodes are homogeneous in terms of attributes values. Then, we apply the (A, R)-*compatible grouping* step which is an iterative step on this partition. After two iterations, we obtain a new partition containing five groups. Only the group A underwent two successive subdivisions. All nodes in the group A_1 interact only with the group B *i.e.* each node of the group A_1 has a relationship with at least one node in the group B. Nodes in the group A_2 have the same list of neighbor groups which are B and C. Finally, nodes in the group A_3 are isolated as they don't have external links. Thus, these three groups are homogeneous in terms of relationships as they have the same list of common neighbor groups.

Similar to k-SNAP method, we propose a two-step method which starts from the *A-compatible grouping*, and then iteratively divides existing groups until the cardinal of the current partition reaches k. However, for the (A, R)-*compatible grouping* step, we propose more refined quality measures. The first measure is a measure of similarity and the second is a measure of separability. On how to divide a group we propose a new mechanism to divide a group based on the notion of *central node*.

5.2   Measure of Similarity Using the Local Degree

The measure of similarity is a combination of two evaluation criteria. The first criterion (see Eq. 10) examines the density of each group in the partition using the local degree; it is a intra-group (IA) criterion. The second criterion (see Eq. 11) evaluates the inter-connectivity of each group in the partition with the other groups; it is a inter-group (IE) criterion.

– Let $P = \{C_1, C_2, ..., C_p\}$ be a partition, $C_i \in P$ be a group and $R_t \in R$ be a relationship type, $IA^t(C_i)$ is defined as follows:

$$IA^t(C_i) = \frac{1}{|C_i|^\gamma} \sum_{x_k \in C_i} Deg_{R_t, P}(x_k) \qquad (10)$$

For a relationship type R_t, it represents the ratio between the sum of the local degrees of all nodes in a group C_i and the cardinal of this group weighted by a parameter $\gamma \in]1, 2]$ for normalization reasons.

– Let $P = \{C_1, C_2, ..., C_p\}$ be a partition, $C_i \in P$ be a group and $R_t \in R$ be a relationship type, $IE^t(C_i)$ is defined as follows:

$$IE^t(C_i) = \frac{1}{|E_t|} \sum_{x_k \in C_i} \overline{Deg}_{R_t, P}(x_k) \qquad (11)$$

For a relationship type R_t, it represents the ratio between the sum of the external edges *i.e.* edge between two nodes not in the same group, and the cardinal of all edges associated to this relationship R_t.

The overall quality measure Δ of a group C_i for a relationship R_t is defined as the ratio of these two criteria. More formally, we define Δ as follows:

$$\Delta = \sum_{i=1}^{|P|} \sum_{R_t \in R} \delta_i^t = \sum_{i=1}^{|P|} \sum_{R_t \in R} \frac{IA^t(C_i)}{IE^t(C_i)} \qquad (12)$$

At each iteration, the aim of this measure is to determine the group that has a weak density and many external links to be divided. Unlike the k-SNAP evaluation, we do not take into account the nodes group membership. We rather consider the nature of links which could be internal or external, regardless the group membership. We select the group C_{i*} and the relationship R_{t*} that make the least local contribution to the value of Δ:

$$(i^*, t^*) = argmin_{1 \leq i \leq |P|, 1 \leq t \leq |R|} \delta_i^t \qquad (13)$$

Once the group C_{i^*} is selected, the mechanism of division consists of determining its *central node*. Then, we divide the group into two subgroups according to the following strategy: one subgroup contains the *central node* with its neighbors, the other the rest of nodes in the group. To better understand this measure, we present an example using the same graph of Fig. 8. Results are illustrated in Fig. 9.

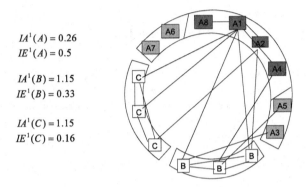

$IA^1(A) = 0.26$
$IE^1(A) = 0.5$

$IA^1(B) = 1.15$
$IE^1(B) = 0.33$

$IA^1(C) = 1.15$
$IE^1(C) = 0.16$

Fig. 9. Graph aggregation using measure of similarity.

As shown in Figure, the output of the *A-compatible grouping* step is a partition composed of three groups in which nodes are homogeneous in terms of attributes values. In the (A, R)-*compatible grouping* step, we compute the quality of each group to determine which will be divided. Setting $\gamma = 1.5$, the intra-group criterion value of the group A is equal to $IA^1(A) = 0.26$ and its inter-group criterion value is equal to $IE^1(A) = 0.5$. For the group B, the intra-group criterion value is equal to $IA^1(B) = 1.15$ and the inter-group criterion value is equal to $IE^1(B) = 0.33$. Finally, for the group C, the intra-group criterion value is equal to $IA^1(C) = 1.15$ and the inter-group criterion value is equal to $IE^1(C) = 0.16$. The quality measure values of groups A, B and C are respectively, $\delta_A = 0.52$, $\delta_B = 3.48$ and $\delta_C = 7.19$. Following these results, the group A is the most heterogeneous because it minimizes the quality measure. As the node $A1$ is the central node of the group, the division is made as shown in (Fig. 9), the first subgroup contains the *central node* $A1$ with its neighbors such as $C_1 = \{A1, A2, A4, A8\}$, the second subgroup contains the rest of nodes such as $C_2 = \{A3, A5, A6, A7\}$.

5.3   Measure of Separability Using the Jaccard Distance

Based on Jaccard distance, this measure evaluates locally the homogeneity of each group in the partition using the Jaccard distance for the different relationships and divides the most heterogeneous group. To evaluate the density of a group, we use the contingency table (Table 5) presented in the previous section for each pair (x_k, x_l) of nodes and each relationship R_t. The quality measure is defined as follows:

$$\Delta(P) = \sum_{R_t \in R} \Delta_t(P) = \sum_{R_t \in R} \sum_{1 \leq i \leq |P|} \delta_i^t \qquad (14)$$

with

$$\delta_i^t = \sum_{x \in C_i} \sum_{y \in C_i} d^t(x_k, x_l) \qquad (15)$$

where $d^t(x_k, x_l)$ is the Jaccard distance (see Eq. 4) according to the relationship R_t.

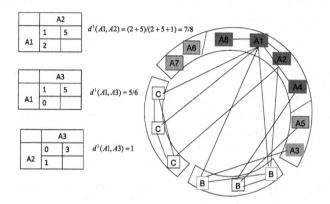

Fig. 10. Graph aggregation using the measure of separability.

To better understand the principle of this distance based on the notion of common neighbor nodes, lets consider again the example of Fig. 8. Results are illustrated in Fig. 10. For the pair (A_1, A_2), the value of the parameter a which represents the number of their common neighbor nodes is equal to 1. Concerning the parameter b which is the number of neighbors of the node A_1 less the common neighbors with the node A_2, its value is equal to 5. Symmetrically, the parameter c which represents the number of neighbors of the node A_2 less the common neighbors with the node A_1 is equal to 1. Thus, the Jaccard distance of the pair (A_1, A_2) is equal to 7/8. Likewise, the parameters a and b for the pair (A_1, A_3) are respectively equal to 1 and 5. However, c is equal to 0 as the node A_1 has one neighbor which is a common neighbor with A_3. Finally, for the last pair (A_2, A_3), the value of the parameter a is zero since they have no common neighbor nodes. In this case, regardless the values of other parameters (b and c) the distance between the two nodes equals 1 thus, the distance is maximum. Using the same procedure, we calculate the remaining distances of the other pairs of nodes. We then evaluate the current partition in order to determine the group to divide. In our case, the group A is the most heterogeneous because it maximizes the quality measure. As the node A_1 is the central node of the group, the division is made as shown in (Fig. 10), the first subgroup contains the *central node A1* with its neighbors such as $C_1 = \{A1, A2, A4, A8\}$, the second subgroup contains the rest of nodes such as $C_2 = \{A3, A5, A6, A7\}$. Note that we obtain the same results with previous measure contrary to k-SNAP measure.

Comparing to k-SNAP, these measures are finer as they are based on the notion of common neighbor nodes regardless the group membership. Two nodes are assigned to the same group according to how they share neighbors. This makes sense when you consider social communities. People who share many friends create a community, and the more friends they have in common, the more intimate is the community. Thus, our objective is to realize a more refined evaluation comparing to k-SNAP. In other words, we no longer based on the notion of common neighbor groups but rather on the notion of common neighbor nodes. The algorithm is summarized below.

5.4   The Algorithm

In this section, we describe our (A, R)-*compatible grouping* algorithm (Algorithm 5) which produces an aggregated graph containing k groups. The inputs of this algorithm are a graph $G = (V, R)$, the size of the target partition k, a subset of attributes $A \subseteq \Lambda(G)$, and a subset of relationships $R - set \subseteq R$ representing user-selected node attributes and relationships.

Algorithm 5. (A, R)-*compatible grouping* algorithm

Input: a graph $G = (V, R)$ where $V = \{x_1, x_2, ..., x_n\}$ is a set of nodes and $R = \{R_1, R_2, ..., R_r\}$ is a set of relationships, the size of the target partition k, a subset of attributes $A \subseteq \Lambda(G)$, and a subset of relationships $R - set \subseteq R$.

Result: aggregated graph containing k groups.

1 $P \leftarrow \{C_1, C_2, ..., C_k\}$ is the $A - compatible$ grouping partition based on attribute values of A.

2 $\Delta \leftarrow 0$;

3 **while** $|P| < K$ **do**

4 $\quad$ **for all** $R_t \in R - set$ **do**

5 $\quad\quad$ **for all** $C_i \in P$ **do**

6 $\quad\quad\quad$ Compute δ_i^t the quality measure of the group C_i for the relationship R_t;

7 $\quad$ Look for $(i^*, t^*) = argmax(min)_{1 \leq i \leq |P|, 1 \leq j \leq |R|} \delta_i^j$ and select the group C_{i^*} to be divided according to the relationship R_{t^*};

8 $\quad$ Look for the central node $x_\alpha \in C_{i^*}$ such that $Deg_{R_{t^*}, P}(v_\alpha) = max_{x_k \in C_{i^*}} Deg_{R_{t^*}, P}(x_k)$ or eventually the central node of the second order $x_\beta \in C_{i^*}$;

9 $\quad$ Keep all nodes $x_k \in N_{R_{t^*}}(x_k) \cup \{v_\alpha\}$ in the group C_{i^*};

10 $\quad$ Put the rest of nodes in a new group $C_{|P|+1}$;

11 $\quad$ update P;

12 $\quad$ update Δ;

Similar to k-SNAP operation, our aggregation operation starts from an *A-compatible grouping* partition based only on attributes and initializes the quality measure Δ to zero (see Algorithm 5 lines $1 - 2$). Then, we launch the *(A, R)-compatible grouping* step which is an iterative step based on the selected relationships until the cardinal of the partition is equal to k (see Algorithm 5 line 3). In each iteration, we have to make the following decisions: (1) how to select a group to divide and (2) how to divide it. According to the chosen quality measure (similarity or separability), the objective is to minimize (respectively maximize) the Δ measure. To do that, we evaluate for the different relationship types the quality measure of each group C_i of the current partition P (see Algorithm 5 lines $4 - 6$). Then, we select the group C_{i*} and the relationship R_{t*} that make the most (respectively less) contribution to Δ value (see Algorithm 5 line 7). This is the *selection phase*. Once the group to be divided is selected, the *division phase* of the group C_{i*} (Algorithm 5 line 8) consists of determining the *central node*, noted v_α, that maximizes its local degree. However, in case that this node has been selected in a previous iteration, we proceed to find the *central node of the second order*. It's the node that has the second highest value of local centrality. The division strategy is performed as follows: the group C_{i*} keeps all nodes $x_k \in N_{R_{t*}}(x_k) \cup \{x_\alpha\}$ (Algorithm 5 line 9) and a new group is generated containing the rest of nodes (Algorithm 5 line 10). At the end, we update the current partition P and the value of Δ and we check if the number of groups is already equal to k (Algorithm 5 lines $11 - 12$).

6   Validation and Experimental Results

In order to evaluate algorithms of *A-compatible grouping* and *(A, R)-compatible grouping* steps, we perform experiments on different types of graphs. First, we study the performance of FCA-RST, K-Medoids and Hierarchical approaches over synthetic heterogeneous graphs. The goal of these experiments is to evaluate the robustness of each approach and its ability to find an *A-compatible grouping* when it is not possible to know *a priori* the number of groups. Second, we realize experiments on a well known homogeneous graph "Network of Political Books"[2] and we compare the obtained results by applying our quality measures with those of k-SNAP. The goal of these experiments is to demonstrate the ability of our aggregation method in producing informative summaries with significant interpretations that could help users in decision-making. All approaches were implemented in Java and the graph data was stored in a GraphML format[3].

6.1   Evaluation of the *A-compatible grouping* Step

To evaluate the FCA-RST, K-Medoids and Hierarchical approaches, three types of heterogeneous graphs (i.e., where nodes are characterized by a different list

---

[2] http://www-personal.umich.edu/~mejn/netdata/.
[3] http://graphml.graphdrawing.org/.

of attributes) were constructed. For the first type of heterogeneous graphs, lets consider Table 3 in which $V = \{x_1, x_2, x_3, x_4, x_5, x_6\}$ is the set of nodes and $A = \{a, b, c, d, e, f\}$ is set of attributes. Using the FCA-RST approach, the A-compatible grouping leads to two groups where nodes in each group have the same list of attributes. This type of graph can be extracted from relational databases where each group represents a different table and nodes are lines in these tables.

The second type of heterogeneous graph is represented in Table 4. Using the K-Medoids approach, the *A-compatible grouping* step gives rise to two groups. The main difference comparing to the previous graph type is that nodes of one group are not characterized by the same list of attributes. However, the similarity in terms of common attributes between two nodes in a same group is bigger than that between two nodes in different groups. This type of graph can represent different objects where some of them have close description.

Finally, the third type of graph is considered in Table 7. It corresponds to the case where the attributes are randomly assigned to nodes. As we can notice, for such type of graph it is harder to find a partition since the similarity between nodes in terms of attributes is almost the same. The goal is to evaluate the robustness of each approach and its ability to find an *A-compatible grouping* partition when it is not possible to know a priori the number of groups.

Table 7. Type 3: random attributes assignment

	a	b	c	d	e	f
x_1	x		x		x	
x_2	x		x	x		
x_3		x	x		x	
x_4	x				x	x
x_5	x	x		x		
x_6		x		x		x

For all approaches and for the different types of graph, we consider the graph shown is Fig. 11 containing 200 nodes and 5970 edges with density equals to 0.3. The set of different attributes in the graph is equal to 10. For the first and the second type of graph, we fix the number k of groups to 2 with an equal size.

Figure 12 shows the results of the *A-compatible grouping* step by applying the FCA-RST approach on the graph presented in Fig. 11. The different pie graphs indicate the proportion of the number of nodes among all groups for the three types of heterogeneous graphs.

As expected, FCA-RST approach can produce two groups of equal size for the first type of graph. However, for the second type of graph, it can not find these groups. The same conclusion for the third type, but in this configuration, the number of found groups is larger than in the second type.

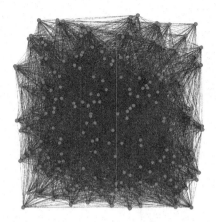

Fig. 11. Graph used in *A-compatible grouping* step for all approaches

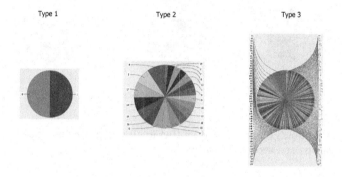

Fig. 12. A-compatible grouping found by FCA-RST approach

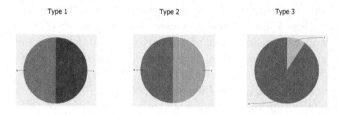

Fig. 13. A-compatible grouping found by K-Medoids approach using Jaccard similarity

Results represented in Fig. 13 show that K-Medoids approach is able to find an *A-compatible grouping* partition containing two groups having the same size for the graph types 1 and 2. However, for the third graph type, the approach found two groups of different sizes. Figure 14 shows results of the same approach K-Medoids but using Jaccard similarity and the simple matching similarity.

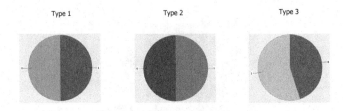

Fig. 14. A-compatible grouping found by K-Medoids approach using Simple Matching similarity

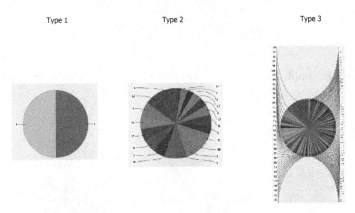

Fig. 15. A-compatible grouping found by Hierarchical approach using Jaccard similarity, average linkage and second derivative

For the two first graph types, the result is similar to the previous one. However, for the third graph type, results are different: the k-Medoids approach using Simple Matching similarity is able to find groups of almost similar sizes.

Figure 15 shows results of the *A-compatible grouping* step by applying the Hierarchical approach using Jaccard similarity to compute the initial matrix, average linkage as measure of similarity between groups and the highest value of second derivative to find the number of groups. The result of the first graph type is equal to the FCA-RST and K-Medoids approaches. Whereas, in the graph of type 2, Hierarchical approach is not able to find a partition of two groups, but it able to generate groups of similar sizes. For the third graph type, a large number of small groups were found. For this approach, it is possible to analyze the corresponding dendrogram of each graph type and use values of similarity to divide it.

Figure 16 shows the dendrogram and the graph using an adequacy cut similarity for the graph of type 2. The dendrogram shows that when the similarity is equals to 0.3, there is two groups.

For the third graph type, the dendrogram represented in Fig. 17 shows very dissimilar groups. The chosen similarity to find the number of groups in the partition is also equals to 0.3. The pie graph shows that there is a large group and

Type 2

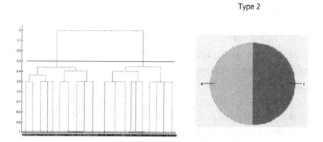

Fig. 16. Hierarchical approach for graph type 2 using Jaccard similarity, average linkage and cut-similarity = 0.3

Type 3

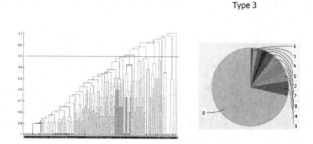

Fig. 17. Hierarchical approach for graph type 3 using Jaccard similarity, average linkage and cut-similarity = 0.3

others 9 small ones. For this graph type, it is a good solution to use dendrogram and similarity to find the better partition.

The results of the experiments for FCA-FRT, K-Medoids and Hierarchical approaches applied to three types of heterogeneous graphs showed that, for all approaches, the graph of type 1 corresponds to an easy problem. This type of graph is characterized by nodes two main different list of attributes, where some nodes have exactly the same list. The second type of graph is a generalization of the first one because there are two main groups of nodes where each group has a very similar list of attributes. In this configuration, FCA-FRT approach could find a summarized graph but with many groups. With the K-Medoids approach, it was easier control the number of groups using the parameter k. Whereas in the Hierarchical approach using the highest value of second derivative to find the number of groups, the result was similar to the FCA-RST method. However, using an adequacy value of similar to find the number of groups, it was possible divide the set of nodes in two groups. Finally, for the graph of type 3, FCA-RST method found a large number of groups. The method based on K-Medoids divided the set of nodes into two groups, since the parameter used is also 2. And for the Hierarchical approach, the use of second derivative showed that this method find large number of groups, but there is the possibility of to cut the dendrogram in specific point to control the summarization.

6.2 Evaluation of the (A, R)-*compatible grouping* Step

In the second series of experiments, we use the real data set "Network of Political Books". This network, compiled by Krebs[4], represents a recent study about books on US politics. Edge connecting books indicate that these latter are purchased by the same customers on the on-line bookseller Amazon.com. Nodes are described by one attribute reflecting its political tendency that could be "liberal", "neutral" or "conservative". These values represent the three modalities of the attribute political tendency. This description was assigned separately by Mark Newman based on reading reviews and descriptions of books published on Amazon. This network is illustrated in Fig. 18, and the corresponding graph contains 105 nodes and 441 edges. Next, we interpret the obtained results by applying our algorithm and the k-SNAP algorithm and make a comparison.

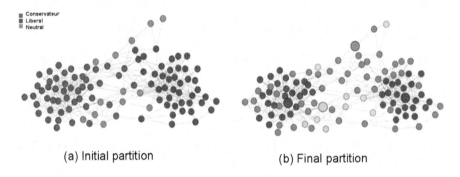

(a) Initial partition   (b) Final partition

Fig. 18. Aggregation using our algorithm (Color figure online).

6.3 Interpretation of Results: Our Method

In practice and for ease of visualization, users are more likely to choose small k values to generate summaries to make a meaningful interpretation. We choose to fix the size of summary graph to 7.

Before engaging in results interpretation, let's recall the principle of our algorithm. It consists of selecting the group that minimizes (or maximizes) the criterion to be divided, then look for the *central node* and divide the group into two subgroups according to the following strategy: one contains the neighborhood of *central node* and the other the rest of the original group.

Applying our algorithm on the "Network of Political Books", the *A-compatible grouping* step generates first a summary formed by three groups in accordance with the modalities liberal, neutral and conservative of the political tendency attribute. As shown in Fig. 18.a, the network is composed of three groups where each node color represents a modality of the attribute. The blue and red groups contain respectively, left-wing and right-wing political tendency

[4] http://www.orgnet.com/divided.html.

books. However, the green group is formed by neutral books. After four iterations, the graph depicted in Fig. 18.b is formed by seven groups. The azure group contains books that are basically neutral but have a conservative aspect. Indeed, each of these books has at least one connection with a book belonging to the red group. Similarly, the green group is formed by books that are neutral but tinged with a liberal character because each of them has at least one connection with a book in the blue group. Finally, the yellow group contains the remaining neutral books from the initial partition. The dark blue and dark red nodes represent respectively, the "most left-wing" and "most right-wing" of existing books. Those who are familiar with current US politics situation won't be surprised to learn that the most left-wing book in this sense was the polemical *"Bushwacked"* by Molly Ivins and Lou Dubose. Perhaps more surprising is the most right-wing book namely, *"A National Party No More"* by Zell Miller. These books correspond to *central nodes* of groups and are represented by a node of larger size. The interest of introducing the concept of the *central node* in the process of aggregation is to summarize each class by a single representing node called also prototype, which greatly facilitates the visualization. Using the *central node* representation, other interpretations can be inferred. We identify in Fig. 19 the existence of two crossing points called bridges that connect liberal books with conservative ones. Also, all books in the yellow group are considered as outliers. This is can be explained by the fact that the intra-group density of this group is zero consequently, it remains stable during the aggregation process and all nodes are seen as central. We can say that these books have little or no influence, and may be considered as noise in data. The identification of bridges and the isolation of outliers are essential for many applications. For instance, the identification of hubs in the Web improves the search for relevant authoritative web pages as they serve as a relay for large amount of information. Furthermore, hubs are considered to play a crucial role in viral marketing.

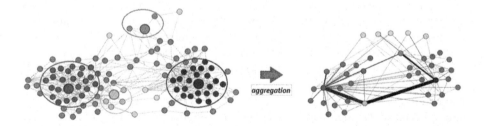

Fig. 19. Aggregation using central nodes representation.

6.4   Interpretation of Results: K-SNAP Method

We use in this last experiment the k-SNAP algorithm to produce summaries from the same data set as shown in Fig. 20. We fix also the value of the summary graph size to 7. After four iterations, the graph is formed by seven groups.

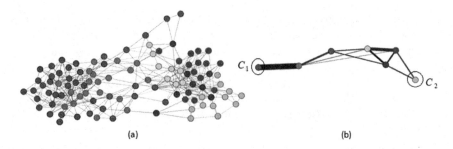

Fig. 20. Aggregation using k-SNAP algorithm (Color figure online).

A visual interpretation shows the existence of two particular groups, C_1 and C_2. As k-SNAP quality measure is based on the notion of neighboring groups, the books belong to these groups are considered as the "most right-wing" and "most left-wing" of the books being studied. Indeed, each book belonging to C_1 (respectively to C_2) is connected only to books in its own community (conservative or liberal). Overall, conservative books show a slightly higher tendency to link with each other than liberal books, which is consistent with the conclusion from the analysis in [29]. Comparing the summaries of the two communities for k = 7, we can see two major differences : the first difference lies in the group of neutral books (green color). This group remains stable throughout the aggregation process. We can say that this group is the most homogeneous one with respect to the quality measure of k-SNAP, which is based on the list of neighboring groups. The majority of these neutral books show a higher tendency to interact with other books from both conservative and liberal groups thus, the summary graph has only one crossing point that connects liberal books to conservative ones. The second difference is that the liberal group underwent three successive divisions : books in this group have very weak connections to other groups but are strongly connected among themselves. According to k-SNAP's algorithm, this group maximizes the quality measure as its nodes are connected to several groups and have not the same list of neighboring groups.

The quality measure of k-SNAP is rather a separability measure, in fact it separates nodes according to the list of neighboring groups. In other words, each node in one group is connected to some node(s) in the neighbor group thus, this quality measure neglects the density of the group (intra-group) and favors the external relations (inter-group). However, our quality measure evaluates locally the homogeneity of each group and tends to divide the least homogeneous according to the principle of *central Node*. The summary graph generates more significant interpretations about the network indicating, as discussed above, the "strength" with which books belong to the communities in which they are.

7   Conclusion and Future Work

In this work, we have presented an original approach for graph aggregation that takes into consideration the heterogeneity aspect of graphs. This approach is soft and adaptive as it allows users to freely select nodes' attributes and relationship types and, to fix *a priori* the size of the summary graph. We have described our *A-compatible grouping* and *(A, R)-compatible grouping* steps specifying formally in each step the proposed approaches. In *A-compatible grouping* step, we have proposed three approaches. The first approach is based on the definition of Rough Set Theory (RST) using Formal Concept Analysis (FCA). The second approach uses the well known K-medoids clustering method. The third one is based on the hierarchical clustering method. For *(A, R)-compatible grouping* step, we have presented two new measures to evaluate the quality of summaries. The first measure is a measure of similarity, and the second one is a measure of separability. To evaluate the performance of our *A-compatible grouping* step, we have conducted experiments on a synthetic dataset including three types of heterogeneous graphs. Results show that our proposed methods are able to find a partition based only on attributes in such graphs. To evaluate the performance of our *(A, R)-compatible grouping* step, we have realized a comparative study versus k-SNAP method. Obtained results demonstrate that interpretations resulting from our method are more significant than those resulting from k-SNAP.

As part of future work, we plan to change the step of settings based on user selection; the goal is not to impose from the beginning a list of attributes, but to dynamically choose the most effective attributes to split the set of nodes. We intend also to make more experiments by evaluating the efficiency and effectiveness of our approach.

References

1. Freeman, L.: A set of measures of centrality based upon betweenness. Sociometry **40**, 35–41 (1977)
2. Girvan, M., Newman, M.E.J.: Community structure in social and biological networks. JProc. Natl. Acad. Sci. USA **99**(12), 7821–7826 (2002)
3. Newman, M., Girvan, M.: Finding and evaluating community structure in networks. Phys. Rev. E. **69**(2), 026113 (2004)
4. Schaeffer, S.A.: Graph clustering. Comput. Sci. Rev. **1**(1), 27–64 (2007)
5. Luxburg, U.: A tutorial on spectral clustering. Stat. Comput. **17**(4), 395–416 (2007)
6. Yan, X., Han, J.: gspan: Graph-based substructure pattern mining. In: ICDM, pp. 721–724(2002)
7. Sun, Y., Aggarwal, C.C., Han, J.: Relation strength-aware clustering of heterogeneous information networks with incomplete attributes. Proc. VLDB Endow. **5**(5), 394–405 (2012)
8. Tian, Y., Hankins, R.A., Pate, l.J.M.: Efficient aggregation for graph summarization. In: SIGMOD, pp. 567–580. ACM (2008)
9. Soussi, R., Aufaure, M.A., Zghal, H.B.: Towards social network extraction using a graph database. In: DBKDA, pp. 28–34. IEEE Computer Society (2010)
10. Santo, F.: Community detection in graphs. Phys. Rep. **486**, 75–174 (2010)
11. MacQueen, J.B.: Some methods for classification and analysis of multivariate observations. In: Cam, L.M.L., Neyman, J. (eds.): Proc. of the Fifth Berkeley Symposium on Mathematical Statistics and Probability, vol. 1, pp. 281–297. University of California Press (1967)

12. Newman, M.E.J.: Detecting community structure in networks. Eur. Phys. J. B **38**, 321–330 (2004)
13. Rodrigues Jr., J.F., Traina, A.J.M., Faloutsos, C., Traina Jr., C.: Supergraph visualization. In: ISM 2006: Proceedings of the Eighth IEEE International Symposium on Multimedia, pp. 227–234. IEEE Computer Society (2006)
14. Shi, J., Malik, J.: Normalized cuts and image segmentation. IEEE TPAMI **22**(8), 888–905 (2000)
15. Ng, A., Jordan, M., Weiss, Y., Dietterich, T., Becker, S., Ghahramani, Z.: Advances in Neural Information Processing Systems. MIT Press, Cambridge (2002)
16. Newman, M.E.J.: The structure and function of complex networks. SIAM Rev. **45**, 167–256 (2003)
17. Chakrabarti, D., Faloutsos, C., Zhan, Y.: Visualization of large networks with min-cut plots, A-plots and R-MAT. Int. J. Hum.-Comput. Stud. **65**(5), 434–445 (2007)
18. Watts, D.J., Strogatz, S.H.: Collective dynamics of 'small-world' networks. Nature **393**(6684), 440–442 (1998)
19. Ren, X., Wang, Y., Yu, X., Yan, J., Chen, Z., Han, J.: Heterogeneous graph-based intent learning with queries, web pages and wikipedia concepts. In: Proceedings of the 7th ACM International Conference on Web Search and Data Mining, pp. 23–32. ACM (2014)
20. Wei, L., Qi, J.J.: Relation between concept lattice reduction and rough set reduction. Knowl.-Based Syst. **23**(8), 934–938 (2010)
21. Shi, C., Niu, Z., Wang, T.: Considering the relationship between RST and FCA. In: WKDD, pp. 224–227. IEEE Computer Society (2010)
22. Stumme, G.: Formal concept analysis. In: Handbook on Ontologies 2009, pp. 177–199 (2009)
23. Jain, A.K., Murty, M.N., Flynn, P.J.: Data clustering: a review. ACM Comput. Surv. **31**(3), 264–323 (1999)
24. Duda, R.O., Stork, D.G., Hart, P.E.: Pattern Classification. Wiley, New York; Chichester (2000)
25. Gan, G., Ma, C., Wu, J.: Data Clustering - Theory, Algorithms, and Applications. SIAM, Philadelphia (2007)
26. Bezdek, J.C.: Pattern Recognition with Fuzzy Objective Function Algorithms. Kluwer Academic Publishers, Norwell (1981)
27. Kaufman, L., Rousseeuw, P.J.: Finding Groups in Data: An Introduction to Cluster Analysis. Wiley, New York (1990)
28. Salvador, S., Chan, P.: Determining the number of clusters/segments in hierarchical clustering/segmentation algorithms. In: ICTAI, pp. 576–584. IEEE Computer Society (2004)
29. Adamic, L.A., Glance, N.: The political blogosphere and the 2004 U.S. election: Divided they blog. In: LinkKDD, pp. 36–43. ACM (2005)

Okkam Synapsis: Connecting Vocabularies Across Systems and Users

Stefano Bortoli[1](✉), Paolo Bouquet[1,2], and Barbara Bazzanella[2]

[1] Okkam SRL, Via Segantini 23, 38121 Trento, Italy
bortoli@okkam.it, bouquet@disi.unitn.it
[2] University of Trento - DISI, Via Sommarive, 14, 38123 Trento, Italy
barbara.bazzanella@unitn.it

Abstract. In the past 10-15 years, a large amount of resources have been devoted to develop highly sophisticated and effective tools for automated and semi-automated schema-vocabulary-ontology matching and alignment. However, very little effort has been made to consolidate the outputs, in particular to share the resulting mappings with the community of researchers and practitioners, support a community-driven revision/evaluation of mappings and make them reusable. Yet, mappings are an extremely valuable asset, as they provide an *integration map* for the web of data and the "glue" for the Global Giant Graph envisaged by Tim Berners-Lee. Aiming at kicking-off a positive endeavor, we have developed *Synapsis*, a platform to support a community-driven lifecycle of contextual mappings across ontologies, vocabularies and schemas. Okkam Synapsis offers utilities to load, create, maintain, annotate, subscribe, and define levels of agreement over user-defined contextual mappings. Furthermore, in order to ease the development of Semantic (Web) applications, Synapsis supports the creation of sets of mappings associated with an application placeholder. On the one hand, this allows developers to easily create and manipulate all the mappings required for their own application without affecting other users. On the other hand, a measure of *Sharedness* for the mappings defined across application contexts is proposed to enable the implementation of ranking metrics that can be used to order the mappings managed through Synapsis. Aiming at supporting a growing number of users, Synapsis was positively tested to be scalable in the order of millions of mappings, performing experiments with synthetic data. Applying the Data-as-a-Service (DaaS) paradigm, the sets of mappings created and managed by Synapsis are also available through REST services, to further facilitate integration into applications working with heterogeneous data.

1 Introduction and Motivation

In the promising vision of the Semantic Web proposed by Tim Berners-Lee [2], the collaborative and distributed creation of semantically annotated documents would enable software agents to perform time-consuming activities on behalf of human users [1]. The community that gathered to realize this ambitious vision

© Springer International Publishing Switzerland 2016
P. Molli et al. (Eds.): SWCS 2013/2014, LNCS 9507, pp. 181–205, 2016.
DOI: 10.1007/978-3-319-32667-2_8

achieved many relevant results with the definition of important standards such as OWL [18,21,26], RDF [23], and the important Linked Data publication principles [3,4]. The combination of these principles with the more recent open data initiative across many countries has been generating a considerable amount of publicly available RDF and OWL data. In recent years, enterprises are attracted by the promise of using such big and rich data to develop new products and services for their customers (see for example [11]). However, exploiting and mining data rise many challenges including the problems of entity matching, ontology matching, and making the data accessible and usable by non-expert users. In the past ten years many efforts were spent in the definition of sophisticated tools for automated ontology matching. These often provided very effective solutions in narrow domains, but a generic automatic reliable solution to the problem is still an open research problem [27]. Furthermore, in [35] it was recently discussed how often even experts have problems in finding agreement on defined ontology mappings. We argue that this is due to two main problems: (1) the intrinsic complexity and heterogeneity of existing ontologies, and (2) the inconsistency and fuzziness in usage of such ontologies caused by contextually interpretable semantics and inherent ontological relativity [28]. Namely, concepts and relations expressed in natural language are interpreted outside the original context of definition and therefore are prone to contextual interpretation. In fact, besides the effort of researchers in formal ontology [15,16,19] the process of ontology definition is driven by specific domain requirements and often ontology engineering practices are neglected [24].

Under these premises, we decided to take one of the ten challenges of ontology matching described in [32] and confirmed unsolved in [27], and propose a novel platform to support a collaborative ontology mappings definition and reuse [37]. The idea to take this challenge is rooted in the pragmatic need of resolving the problem of semantic heterogeneity affecting a knowledge-based solution of the entity matching in the context of the Semantic Web [5]. In particular, in this work we argue that collecting and maintaining ontology mappings as contextual bridge rules [8] to harmonize the semantic of entities' attributes can provide great benefits by enabling the application of knowledge-based solutions to an entity matching problem [5]. Therefore, in our attempt to solve the entity matching problem in the linked data, we produced several thousands of mappings from existing ontologies, schemas and vocabularies towards a target ontology named Identification Ontology[1] ([6]). Often these mappings were produced without considering the original, or intended, semantics of the properties, but rather relying on its actual function looking directly into the data. This approach, besides being practical and concrete, interprets the ontology mappings as contextual analogies as suggested in [29]. Namely, when producing mappings, rather than considering the similarity among originally intended functional purpose of the properties (homology), we consider also its real function (analogy) so that the mapping relation holds primarily on the instance level. On the one hand, we are aware that this approach will create mappings that might not be absolutely coherent

---

[1] http://models.okkam.org/identification_ontology.owl.

and correct across several contexts, but as long as they serve the purpose we can live with this limitation. On the other hand, we want to use a first core set of mappings to kick-off a positive endeavor for the definition of a platform to support a community-driven lifecycle of contextual mappings between ontologies, vocabularies and schemas that could serve the definition of new applications exploiting open linked data.

In this work we describe Okkam Synapsis, a web application conceived to support the linked data community in creating, sharing and reusing contextual ontology mappings to support the development of novel services based on the linked data consumption. Okkam Synapsis offers utilities to load, create, maintain, annotate, subscribe, and manage levels of agreement over user-defined contextual mappings. Most importantly, endorsing the recommendations described in [35], we support different fine-grained typing models for the definition of mappings (e.g. OWL and SKOS) and compute level of agreement according to different measures to support the mapping ranking. The mappings produced will be available also through REST services, providing several levels of selection to support diverse and unforeseen application scenarios. The purpose of the application is to enable the users of Okkam Synapsis to collaborate in the definition of mappings, commenting, rating, and subscribing them. Furthermore, we want to allow users to explicitly define the context of use of the defined mappings (i.e. the application scenario), so that other users can take informed decision about reusing mappings.

The underlying assumptions are:

- real linked data is in general too messy to rely on a unique set of mappings in different contexts of use;
- linked data may change across time, therefore contextual mappings must be subject to specific lifecycle;
- the number of existing vocabularies is growing, but reuse practices make the manual mapping process feasible (see Linked Open Vocabulary[2]);
- perfect agreement about defined mappings is unlikely to happen [35], it is better to let users to select what they need;
- the level of agreement on mappings across contexts is a promising parameter for the implementation of filtering and ranking procedures.

The focus of this work is on proposing a clear position about the creation and management of ontology/schema mapping, interpreting them as contextual bridge-rules regardless the original intended meaning (or semantic) of the involved concepts. We focused on defining a user interface suitable to support daily activities of linked data practitioners (as we are among them), adopting and adapting it according to common sense usability requirements of the developers involved. We did it in the pure spirit of creating a comfortable workbench to define a truly collaborative space for semantic web practitioners. The core of the proposal is on describing a collaborative space suitable to scale up to manage a possibly diverging amount of mappings, whereas evaluating experimentally

---

[2] http://lov.okfn.org/dataset/lov/.

the user satisfaction is postponed to the near future. The reminder of the paper is organized as follows: in Sect. 2 we overview the related works dealing with crowd-sourcing of ontology mappings and other community-driven approaches; in Sect. 3 we present in detail the vision and the role we foresee for Synapsis; in Sects. 4 and 6 we describe in detail the platform, discussing functions and services. In Sect. 5 we discuss a measure of *Sharedness* to implement ranking functions for mappings; Sect. 6.1 presents the results of an initial analysis of the scalability of the platform; Sect. 7 describes the kick-off dataset of mappings; and finally in Sect. 8 we describe future work and outline some concluding remarks.

2   Related Work

According to the most recent survey [27], there are not many tools support-ing collaborative creation of ontology mappings. A first attempt of extending the notion of ontology matching to community-driven ontology matching can be found in [37]. The authors propose a community-driven approach which involves user communities in the processes of evaluating mappings and reusing them based on the user profiles and their community and social relationships. For the purposes of the present work, the paper introduces the interesting notion of "subjective alignments" which is in line with our idea that mappings have subjective or contextual relevance (i.e. mappings which are appropriate for spe-cific tasks in a specific community, may be inappropriate or even contradicting to practices of other communities). As a consequence, the reuse of mappings created by different users implies resolving issues such as the appropriateness of mappings when using them in different contexts, trust issues and measures of agreement. The proposed solution in [37] is based on the creation of anno-tated mappings to represent information such as who created the mapping or usage-related characteristics, which may be used to select the most appropriate mapping for a certain task or user. In this work it is proposed a system where the mappings are entirely defined by the users, and measures are introduced to support efficient ex-post reuse. Whereas the community-driven ontology match-ing approach proposes a semi-automatic system where users provide feedback on automatically generated mappings and save mappings for reuse in annotated form. Even though the approach presents interesting insights and has been tested with a set of initial experiments, the solution seems not to be developed beyond the prototype stage and has not been made publicly available. In addition, to the best of our knowledge, the scalability of the system has not been tested.

Extending the concept of annotated mappings, Noy et al. [25] proposed an extensible annotation model to represent community-based mappings that con-tains a metadata model to describe mappings. The model was validated by extending a repository of biomedical ontologies called BioPortal with a user interface to create, search, filter, visualize and comment mappings. More than 30000 mappings have been collected from 7 sources. Currently the system is up and running and it stores 445 ontologies with more than 3 million concepts. The BioPortal mapping system presents several aspects in common with Synapsis, but some important differences need to be pointed out. First of all, both systems focus on mappings between individual concepts rather than ontologies, that is mappings are first-class objects in the corresponding repositories and they can be created, reused, commented or discussed by the community of users. In both systems mappings are annotated with metadata to describe the mapping features (e.g. author, creation date) and facilitate the reuse. Both BioPortal and Synapsis allow to specify equivalence, similarity and subsumption mapping relationships and both implement utilities to enable users to reach consensus on mappings (for example by commenting mappings and replying to comments). In addition BioPortal supports the visualization of mappings and related features. However the two systems have a very different focus and breadth. BioPortal is a domain-specific system service which is proposed as a comprehensive repository of biomedical ontologies in which mappings are part of a rich set of metadata provided by the system to disclose the value of such resources. Mappings are not the first business of BioPortal which uses them to augment the services on top of its repository such as navigation or access to biomedical resources. On the contrary Synapsis is domain neutral and it is not envisioned as an ontology repository but as a community-based system for creating and sharing ontology mappings covering any sort of ontology, vocabulary or metadata schema from any domain or community of users. Another important difference is on the source of mappings. BioPortal mappings come from different sources (using different mapping approaches or algorithms) since the system is not designed to be a primary environment for creating large volumes of mappings but as a system to upload and share mappings generated with external tools. Only a small set of mappings in BioPortal are mappings generated by users through the Web user interface of the system. Instead, Synapsis offers a platform whose main purpose is to create and manage mappings and the main source of mappings is user-generated mappings within the Synapsis environment. Finally, while Synapsis returns a ranked list of mapping results for each mapped concept which is based on many possible indicators (including sharedness measure described in Sect. 5), the BioPortal system simply lists the candidate mappings without considering any indicator that could help easing the reuse. An overview of the comparison between BioPortal and Synapsis is shown in Table 1.

To the best of our knowledge, there are no other open systems currently available to create, share and manage community-based ontology mappings even though other initiatives have investigated the role of human contribution in the ontology alignment problem. In [30] it is described CrowdMap, a solution for ontology matching based on crowd-sourcing. The ontology matching task is decomposed in micro-tasks and submitted to workers of crowd-sourcing platforms such as CrowdFlower and MTurk for manual evaluation. The results of an initial set of experiments show that CrowdMap on average is able to outperform automatic alignment solutions, providing a first evidence that the combination of ontology matching algorithms with human-driven approaches may increase the quality of the mapping work flow. However, compared to the pragmatic approach that we propose in this paper where users create the mappings based on

Table 1. Comparison between BioPortal and OKKAM Synapsis

	BioPortal	Synapsis
Domain	biomedical	any
Purpose	ontology repository	mapping management
Mapping metadata	yes	yes
Mapping visualization	yes	yes
Mapping edit	no	yes
Mapping main source	external	internal
Mapping generation	automatic, user-defined	user-defined
Mapping relations	OWL, SKOS	OWL, SKOS
Relevance metrics	no	yes
Support applications	no	yes
Mapping properties	no	yes
Consider context path	no	yes
Tree view of ontology	no	yes
Main Target	bio/medical research	linked data
Scalability	with 30k mappings is slow	tested with 100M
Online accessibility	bioportal.bioontology.org	api.okkam.org/synapsis

their real needs and contextual tasks, the CrowdMap approach has the disadvantage to assign the mapping tasks to people which are not supposed to use the mappings in real applications or systems. In addition the typical methodology of the crowd-sourcing, where tasks are published on specialized Web platforms where workers can pick their preferred tasks on a first-come-first-served basis can not guarantee that tasks are performed by workers with an adequate level of domain knowledge or expertise. In addition, since concepts in micro-tasks are under-defined and there may not be a unique answer to a matching micro-task, the approach has the limit of not providing a minimum level of contextual information which can help the workers in performing the mapping task reducing the uncertainty. A solution to address the uncertainty issue in crowd-sourcing ontology matching is proposed in [12] where the authors investigate different ways to design micro-tasks and ask for the workers contribution. Since the work is at an embryonal stage, it is hard to evaluate the feasibility of the approach. In our proposal, we aim at tackling part of the issues related to uncertainty of the mapping definition by allowing to define a specific context path which provides additional information to interpret the mapping. Finally the micro-tasks approach, as implemented in [30], does not address a way to use agreement among workers to determine the certainty of mappings.

The use of collaborative features to achieve agreement on mappings of concepts and facilitate their reuse is a cornerstone of our community-based approach and has been investigated by Correndo and Alani in [17]. The authors propose

OntoMediate, a prototype system which provides a collaborative solution to address the task of aligning ontologies, by extending and enhancing automatic mapping tools with community support. In OntoMediate users can collaborate on aligning their ontologies, and manually-driven alignments can be stored and reused later. Like in Synapsis, mappings are seen as a resource built and shared through the system, but the main goal of the OntoMediate collaborative environment is to extend and enhance automatic tools rather than to provide a primary environment for mapping creation and curation. The system, developed within the UK OntoMediate project, is currently not available for public use. In [11] the authors describe Helix as a tool for creating ontology mapping as a pay-as-you-go task while consuming linked data. An interesting aspect of this work is the use of an interface which aims to engage users progressively in the process of ontology alignment through the course of their everyday use of the system. Unfortunately the system is not publicly available online for further analysis. Another trend in managing collective ontology matching is through gamification. In [24,33] are described Guess What?! and SpotTheLink proposing the solution of ontology matching tasks in form of games to give incentives and foster engagement to ease the cognitive effort of users and stimulate the creation of mappings and links in the linked data cloud. Noticeably, to the best of our knowledge these systems are not currently available. In this context we do not consider papers presenting automatic solutions to the ontology matching for which we refer to the aforementioned survey [27].

In light of the analysis presented, the only system available providing services comparable with the one of Okkam Synapsis is BioPortal [25]. However, given the domain specific purpose of BioPortal and the identified differences, we can safely affirm that there is room for a solution such as the one proposed in this paper.

3 Vision and Perspectives: Collaborative-as-you-go

The kick-starter reason for the conception of Synapsis as a platform for the collaborative management of schema and vocabulary mappings is the need of sharing the burden for the creation and maintenance of mappings themselves. In fact, the creation of valuable mappings requires a substantial amount of human effort and domain knowledge either to create them manually or to configure (e.g. creation of training set and tuning) the application of the sophisticated algorithms defined so far. The usage of automated tools is convenient just if the ontologies are large enough to justify the effort of their contextual configuration. This usually happens in medical science and biology domains. Whereas, because of the relatively small size of average ontologies available online, most people rely on the definition of purposely defined mappings, even relying on crowd-sourcing platforms (e.g. [30]), that are bound to the application in which they are used. Through Okkam Synapsis, we aim to enable semantic application developers in sharing and reusing ontology mappings (manually or automatically defined) to create a positive endeavor that would speed up the work of the most by sharing the effort of the creation. As mentioned in the introduction, mappings

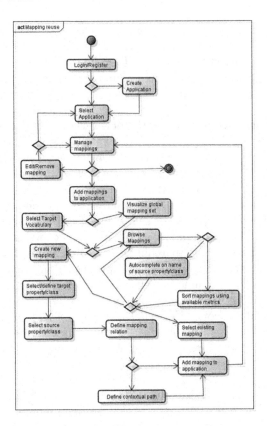

Fig. 1. Mapping reuse activity diagram

are often contextualized to satisfy specific application requirements. This is very likely to create a long tail of not reusable mappings that can make the reuse of mappings a cumbersome task. However, leveraging defined contexts enables the estimate of how many times a mapping is reused (considering different degrees of compatibility). This allows the measurement of the level of agreement around each single mapping, and therefore provides tools to reduce the negative effect of the long-tail affecting the reuse of mappings. Therefore, through Synapsis, developers can create a set of mappings for their application by reusing and filtering among existing one. In the context of the application, developers can manipulate and filter mappings according to their specific needs and make sure that those mappings will always be available for his application. A graphical representation of the activities a user is expected to carry out using Okkam Synapsis is presented in the activity diagram depicted in Fig. 1. In satisfying this *private need*, the developer helps Synapsis in estimating the level of re-usability of the mappings, and therefore potentially ease the job for the next developers. We define this a *collaborative-as-you-go approach*, as the collaborative effort comes from the direct satisfaction of a private need in a public environment, without the explicit will of 'donating' time. It is important to notice that openly free mapping donation is still possible, as people can still share mappings in the best collaborative spirit. To support also these types of users, we plan to extend Synapsis with features to assign badges and scores to the most active collaborators. The main challenges in the bootstrap phase of this vision are the creation of a first relevant set of mappings, and the definition of a user interface supporting an efficient human-computer interaction that makes the reuse more effective than the bare creation of the mappings from scratch. In this paper, we attempt to deal with both of these issues, presenting in detail the structure of the mapping and all the meta-data foreseen to be useful for their selection and filtering to facilitate mappings reuse. Furthermore, Synapsis allows to import and export sets of mappings supporting known formats (e.g. Alignment API [10]), and we aim to define a set of web-services supporting the retrieval of mappings according to the data-as-a-service paradigm. The first real-world application of Synapsis is related to the management of a set of mappings used to support the knowledge based solution for entity matching defined in [5] and currently part of the Okkam Entity Name System [7]. We claim that the collection and management of ontology mappings in a public space will generate further benefits beyond the reuse, including benchmark for ontology mapping tools, and ground to experiment and evaluate inter-rater agreement metrics based on regression techniques. It is important to underline that the present work does not deal with any of these specific challenges, but mostly argues in favor of the adoption of the Okkam Synapsis to ease the creation of applications relying on linked data.

3.1   Use Case Example: Reusing Open Linked Data

In order to clarify the use of Okkam Synapsis, in this section we outline a use case where a developer wants to reuse Open Linked Data with a private ontology. Applying basic ontology engineering guidelines, the developer builds its own domain ontology, combining properties and concepts from common ontologies (FOAF, DBPedia, Schema.org) to the one defined for the specific domain. After the definition of the model, the developer collects some datasets performing queries to DBPedia, Freebase, and few other open dataset interesting for the application domain. After analyzing and performing some basic cleaning operations, the developer is well aware of both the domain ontology and also how the data look like. Hence, logs in Okkam Synapsis and creates a new mapping set by registering the new application. Then, following the activity diagram depicted in Fig. 1, the developer can browse the existing mappings, selecting the ones to be reused (e.g. dbpedia:Person is equivalent to foaf:Person), and creating the one that are missing. When the task is complete, the developer has several options to use the defined mappings' set. These can be downloaded selecting one among the available formats (e.g. JSON, or Alignment API [10]), or can be integrated within the application logic using the Okkam Synapsis REST APIs. Within Okkam, we use Okkam Synapsis to manage mappings between defined application ontologies and the Okkam Identification ontology [6], currently used to define a knowledge based entity matching solution for the Okkam Entity Name

System. In particular, the mappings are applied along reconciliation tasks executed using a customized version of Open Refine[3] [36]. In fact, when we want to 'okkamize' a dataset (i.e. assign an ENS id to the entities mentioned), we first load the dataset into Open Refine and perform some cleansing operations (e.g. uniforming date representation). When the data is ready, a custom component of Open Refine queries Okkam Synapsis to download the mappings and uses them to rename the Open Refine columns. It is important to notice that the same column could be mapped to different properties of the target ontology according to the specific application needs.

Table 2. Example of dataset to be mapped

foaf:name	foaf:phone	dbpedia:city	dbpedia:state	dbpedia:country
John Smith	123456	London	Ontario	Canada
Bob Black	321654	London	Kentucky	USA

For example, consider Table 2, where two fictional records about people are represented. In a first stage, the location is considered as the primary entity and the last three columns are used, mapping *dbpedia:city*, *dbpedia:state* and *dbpedia:country* to properties related to properties of the Identification Ontology, therefore using the mappings (*dbpedia:city* → *id:location-name*),(*dbpedia:state* → *id:first-order-administrative-division*),(*dbpedia:country* → *id:country*). In a second stage, when the person is considered the primary entity, another set of mappings is used: (*foaf:name* → *id:name*),(*foaf:phone* → *id:phone-nr*), (*dbpedia:city* → *id:city-of-residence*). Because the city of residence was disambiguated and reconciled, the reference to the reconciled id it sufficient now. Once the identity of the entities involved is disambiguated, a third set of mappings can be used to transform the data using the application ontology defined. The proposed example has the twofold purpose of clarifying how mappings become an essential part of data manipulation in applications using semantic data and how different mappings are required to interpret the semantics of attributes within specific application constraints.

4   User Interface and Features

The current version of Synapsis distinguishes between three kinds of users: administrators, end-users, and anonymous users. Administrators are users that have unrestricted access to all the user-level functions, including uploading a source ontology, creating new mappings for concepts and properties, deleting existing mappings, setting/changing the status of defined mappings, evaluating existing mappings and reusing/exporting mappings. End users have only access to social functions to express their level of agreement on previously created mappings and reusing them. They can endorse and comment existing mappings,

---

[3] http://openrefine.org/.

follow mappings they are interested in, rate mappings and export mappings. End users can crate applications, and these contexts, manipulate and change existing mappings. All operations executed in the application context will be available also to others, but will not affect directly others' applications. Anonymous users can browse, select and export sets of mappings. Figure 2 shows a snapshot of the User Interface of Okkam Synapsis which presents three main areas: the (target) ontology on the left, the mappings in the central part and the mapping filters on the right. The user can select one of the ontologies/vocabularies currently present in the platform from the drop-down menu on the top-left corner of the page or import a new ontology selecting the Import function from the Function button. Following [25], we call the selected/uploaded ontology the *Target Ontology*, which is the ontology whose concepts/properties the user wants to map. According to this naming convention, a mapping can be seen as a relationship between two concepts/properties in different ontologies. Each mapping has a source concept/property, a target concept/property, and a mapping relationship which connects the source concept to the target concept. After having selected it, the target ontology is loaded, processed and represented as an indented tree on the left side of the interface. The choice of using an indented tree is based on the study described in [14], where users evaluated this representation model as easier to use and more understandable than alternative models such as the graphs. With the primary objective of enabling users in defining mappings, we decided to flatten the ontology to a list of concepts and present the properties attached to them. In the current version, we rely on a simple RDF processor implemented relying on Apache Jena API[4]. The selection of a node of the target

Fig. 2. Synapsis user interface

---

[4] https://jena.apache.org/.

ontology triggers the loading of all the mappings defined for that concept or property in the central part of the window. Furthermore, Synapsis supports the creation of mappings even if there is no loadable ontology, as in the case of `schema.org`, or metadata schemas like VIAF[5]. Each mapping is composed by the following features:

- Source and a Target URI: we call Source the URI of the resource (i.e. source concept) mapped towards the element of the target ontology or a generic Target URI. Notice that mappings in this context are preliminarily interpreted as directional, neglecting reflexivity and other meta-properties of the used relations.
- Relation Type: the type of relation between the Resource URI and the Target URI. The user can select among a number of relation types including OWL meta-relations such as `owl:EquivalentProperty`, `owl:EquivalentClass`, `owl:SubClass`, `owl:SubProperty` and SKOS meta-relations `skos:exact`, `skos:close`, `skos:broader`, `skos:narrower`. In this context, we neglect `skos:related` and `skos:unrelated` because we believe these types of relations are of little interest. In order to help the user in choosing the right type of relation we refer to the guidelines proposed in [35] and still available at [34] as appendix A2. It is important to underline that, in this version, we do not perform any kind of inference relying on the properties and related entailment of the chosen mapping relations.
- Contextual Path: this element aims at providing unambiguous reference to a specific ontological context in which the source URIs should be interpreted. In fact, many properties within the hierarchy of a model can have the same URI but different meanings when associated to different concepts (or Domain) (similar issue has been addressed by Serafini et al. in [31]). For example, consider the attribute *title* for a movie and a person. Syntactically, the URI is the same, but the semantic is different. Therefore, not considering the domain of interpretation of a property (i.e. the context) may lead to erroneous mappings selection. This type of problems is even more evident when models heavily rely on concepts and object properties rather than using data properties. This may require the definition of paths to exhaustively disambiguate a property domain.
- Status: a label among `Raw`, `Edited`, `Closed`, `Accepted`, declaring the status of a mapping. These labels are assigned by administrators of Okkam Synapsis keeping into consideration time and opinions expressed by the members of the community.
- Author: the author of the mapping.
- Description: A description of the resource mapped possibly coming from official documentation.
- Contextual Sharedness: Each mapping presents a score based on the measure defined in Sect. 5.1 that can be used to sort and filter mappings.
- Number of Watchers: any mapping can be watched by a member of the community. Watching a mapping allows users to be notified about activities concerning the mapping.

---

[5] Virtual International Authority File https://viaf.org/.

- Number of Likes: any mapping can be *liked* by a member of the community. A like essentially implies an agreement and a subscription to possible events related to the mapping.
- Comments: members of the community are enabled in commenting and discussing about a mapping. We foresee cases where people may ask for clarifications and argue about the validity of the mapping.
- Contextual Tags: any mapping is annotated with a set of tags which identify fuzzy contexts of application of the mappings. These tags can be used to search and filter mappings.
- Time since creation: any mapping is annotated with a fuzzy representation of the age of the mapping based on the creation time.

In Fig. 2, one can see the list of all the mappings about the OCCUPATION property of the concept *Person* defined in the Identification Ontology. Each mapping is represented by a source URI preceded by two graphical symbols describing respectively the relation type and the status of the mapping. On the right of the source URI, one can see the author of the mapping and whether the mapping was watched and by how many users. Finally, we show the number of people care about that specific mapping. Then, on the right side of the interface, a set of filtering options are provided. The user can filter mappings according to these main dimensions, and typing on the top input field, can filter mappings relying on the name-space or the local part of the mappings URIs. The creation of the mappings relying on an imported target ontology is not the only way supported by Synapsis. In fact, on the top tab, it is possible to select the global view or the application view. The tabs provide different perspectives on the set of mappings managed by Synapsis. A view of the application view is presented in Fig. 3. The Application view shows just the mappings that were created for a specific application (e.g. "My application" in Fig. 3). The main way of interaction we foresee, is the one where in a first phase the user navigates, filters and selects a set of mappings in the application and global view relying on auto-complete and other filtering features available. The collected mapping are marked to be managed in a specific application by triggering an action on the button right above the *Relation* field presented in Fig. 2. Then, when this first collection phase is completed, the user can move on the application part, and perform all the necessary edits, adding tags, comments and anything that is required for the specific application purposes.

Clicking on each mapping, the user can visualize all the details about the mapping in a specific detail page (as shown in Fig. 4). This page allows to add comments, rate, subscribe and add possible contextual tags in a collaborative manner. Ratings are made on a 6-item scale including the following options: approved (i.e. the source and target concepts both mean the same thing), broader (the target concept should be a broader term than the source concept), narrower (i.e. the target concept should be a more specific term than the source concept), related (i.e. the two concepts are not an exact match but they are closely related), not sure (i.e. there is a relationship between the two concepts but none of the above relations are appropriate or the term is used in a confusing or contradictory

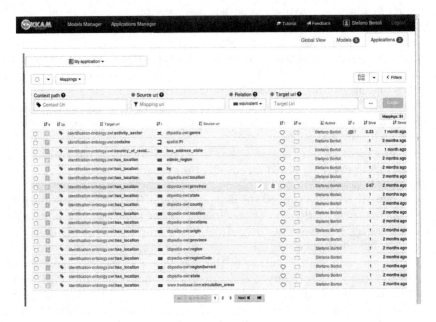

Fig. 3. Okkam Synapsis user application view

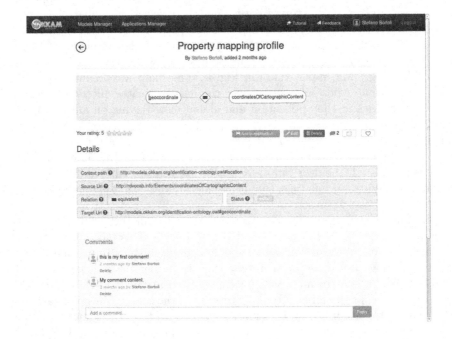

Fig. 4. Mapping profile page

fashion), rejected (i.e. the two concepts are definitely not the same, nor do they have any other direct relationship with each other as listed above). The mapping detail page essentially aims to provide tools for the collaborative interaction for each single defined mapping. If a user subscribes a mapping, any notification will include a link to the specific mapping detail page.

Once selected the target ontology, the user is enabled in filtering mappings according to different features such as author, threshold on sharedness measures, etc. (see Fig. 5). On the right part of the page, the filter features are displayed, and the user is enabled in selecting them. Each selection triggers an action on the list of mappings, removing the filtered ones. Mappings can be filtered by author name, status, relation type, rating and creation date. It is also possible to select all the mappings that have comments.

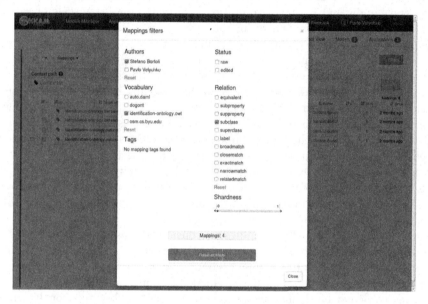

Fig. 5. Mapping filtering options

5 . Measuring Cross Applications Agreement

In [35], the Tordai et. al propose an overview of methods to measure agreement among experts in evaluating mapping across different vocabularies. To accomplish this, the authors relied on four statistical tools used in different fields to measure inter-rater agreement:

- *Observed agreement*: is computed by summing the elements on which rater agree, divided by the total number of elements rated.
- *Cohen's kappa* [9]: is an elaboration of observed agreement, considering the possibility that two raters agree by chance. Therefore *Cohen's kappa* is a more conservative metric, as agreement tends to decrease faster than in observed

agreement estimation. This metric works to estimate agreement between two raters, and therefore, the usage of this metric requires to compute an aggregate average [35]. Interestingly, this metric can also be associated to weights to enable agreement estimation among similar, but not equivalent categories (e.g. exactMatch and closeMatch).

- *Fleiss' kappa* [13]: is related to Cohen's kappa statistics, but allows to estimate agreement across multiple raters without assuming all the elements were rated by everyone.
- *Krippendorff's alpha*: is a more versatile inter-rater reliability metric based on estimation of 'disagreement', supporting evaluation even with missing data [35].

All the considered metrics produce statistical estimations of inter-rater agreement, however their interpretation is not obvious. In [35] the authors propose to rely on Landis and Koch interpretation [22], defining a fuzzy scale of values to different intervals of agreement among raters (e.g. between 1.0 and 0.8 there is *almost perfect agreement*, between 0.8 and 0.6 there is *substantial agreement*, etc.). However, in the context of Synapsis, the approach must be considered carefully. In fact, Synapsis's users are not asked to rate/classify anything explicitly and *per se*, but they rather work targeting private application objectives and consider mappings as "analogies" (as suggested in [29]). So, in this context we could interpret reuse as-is of mappings across applications as if two users would agree on the creation of the same mappings. However, this distinction would not be completely representative of the settings. In fact, it is important to stress the fact that mappings are not defined "per se", but they are rather defined as tools to be used on real data considering all possible inconsistencies. Therefore, the same user could map differently the same resources in different application scenarios. This moves the focus of agreement estimation to the application, or context, rather than the user. Mappings are considered pragmatically as "doings" (or deeds) instead of "abstract" relations between static vocabularies. We find this approach coherent with the findings described in [20], where "knowing what they would be used for" would have helped raters in evaluating whether they would agree or not on defined instance mappings. Therefore, our conclusion is that inter-rater agreement metrics are not the right tool for our purpose, and we need to define a different metric.

Therefore we introduce a new measure that helps evaluating the level of agreement across contexts for the mappings defined in Synapsis. We name this measure *Contextual Sharedness*, interpreting applications as contexts of reuse of the mappings, and thus the primary source of sharedness estimation. Our aim is to use this measure as a first ranking measure to order the potential candidates for a given target concept, and in the long run, marginalize mappings with a low degree of sharedness. It is important to understand that an in depth analysis of the impact of such measure is out of the scope for this work, and will rather be the focus of future experiments.

5.1   Measuring the Contextual Sharedness in Synapsis

A mapping in Synapsis is conceived as a relationship between two concepts in two different models (e.g. ontologies). For the purpose of this discussion, we define a concept (C) as a *conceptual element* of the model which can be a class or a property. Each mapping defines a directional relation connecting a source concept C_s to a target concept C_t within a given context (or application). Formally, a mapping can be represented as a triple: $M : \langle C_s, C_t, R_M \rangle$ where C_s is the source concept, C_t is the target concept, R_M is the mapping relation which connect C_s to C_t. Note that in Synapsis different R_M can be specified such as *owl:equivalentProperty, skos:exactMatch, skos:closeMatch*, and others and therefore different mapping relations between C_s and C_t can be represented. We treat these mappings as distinct. This means that we can compute a different contextual sharedness for each R_M linking C_s with C_t. Moreover, in future we can introduce other measures to establish the most "agreed" relation among those defined for each pair of concepts.

Our aim is to define a measure, called *Contextual Sharedness CS* which measures how much a certain mapping is shared across Synapsis contexts and therefore it can be suggested to be re-used in a new context. Intuitively, the idea is that, if many contexts share the same mapping between a source concept and a target concept and in only few contexts a different source concept is put in relation with the target concept (or a different R_M is used to connect the source and the target), then the degree of association between source and target should be high, indicating that there is a high probability that the considered mapping is useful in a future context. Therefore, a high ranking position should be assigned to the mapping. To translate this intuition into a quantitative measure, we define the *Contextual Sharedness* for a given mapping M as the ratio between the number of contexts in which $C_s \in M$ is mapped onto $C_t \in M$ through $R_M \in M$ and the number of contexts in which any C_s is mapped onto $C_t \in M$ through whatever relation. Formally, we define the *Contextual Sharedness* as follows. Let's represent the Target Ontology as a vector $T = (t_1, ..., t_i)$ of target concepts and let's define the Source space for T as the set of source objects which are mapped to some $t \in T$ through any relation R. The Source Space can be represented as a vector $S = (s_1, ..., s_j)$. We can represent the contextual mapping domain for T under a given R, as a frequency matrix $MR = [mr_{ij}]$ where $mr_{ij} \in \mathbb{R}^+$ counts the number of contexts which map s_j to t_i through R. If $mr_{ij} = 0$, this indicates that there is no context in Synapsis which maps, with the relation R, s_j to t_i and therefore s_j and t_i are unrelated in that representation domain according to R. We finally define a frequency matrix $M = [m_{ij}]$ where $m_{ij} \in \mathbb{R}^+$ counts the number of contexts which map s_j to t_i through any mapping relation. The only difference between MR and M is that MR denotes the degree of association (computed as the number of contexts which share the mapping) between the source concept and the target concept according to a specific type of mapping relation, while M measures the association independently by the type of mapping relation. We define the *Contextual Sharedness CS* for a target ontology T according a certain relation R, as a matrix $(i \times j)$:

$$CS_{ij} = \left[\frac{mr_{ij}}{m_{i.}} \right] = \left[\frac{mr_{ij}}{\sum_1^j m_{ij}} \right]$$

The Contextual Sharedness is a measure that ranges between 0 and 1 ($0 < cs_{ij} < 1$). cs_{ij} is equal to 1 when in all contexts represented in Synapsis the source concept s_j is mapped to t_i according to R, while it is equal to 0 when there is no context in Synapsis mapping s_j to t_i using R as mapping relation. This can happen when s_j is unrelated to t_i but also when there are contexts which map s_j to t_i using a mapping relation different from R. In Synapsis the CS measure has been implemented as a ranking index to order the source concepts given a target concept. When the user selects a target concept (for example to enter her/his mapping), Synapsis returns the list of candidate concepts ranking them based on the CS measure.

To make a concrete example of the use of the CS measure, imagine that a user of Synapsis, that wants to use the dbpedia ontology in a certain application, for the first time creates a mapping between the C_t *dbpedia:Country*[6] and C_s1 *schema:AdministrativeArea*[7] through a `skos:exactMatch` mapping relation. As long as no other Synapsis user creates a mapping between C_t and some other concept or between C_t and C_s1 through a different mapping relation, the CS measure which links C_t and C_s1 is 1. This means that the only mapping suggested by Synapsis for *dbpedia:Country* is a `skos:exactMatch` to *schema:AdministrativeArea* with $CS = 1$. Now assume that another user adds a different mapping to the target concept, C_t, *dbpedia:Country*, in a different context, let's say he maps the C_s2 *linkedgeodata:Country*[8] to the C_t *dbpedia:Country* through the `skos:closeMatch`. Now the CS which measures the association between C_t and C_s1 for the relation `skos:exactMatch` is $1/2 = 0.5$. As a consequence when a third user wants to create a mapping for C_t the systems will return two potential candidates, with the same probability (or CS), i.e. a `skos:exactMatch` to *schema:AdministrativeArea* and a `skos:closeMatch` to *linkedgeodata:Country*. If the third user chooses to map C_t to *schema:AdministrativeArea* through a different relation, the CS between the two concepts will decrease to 0.33 while if the same relation as that used by the first user is chosen the CS will increase to 0.66. In the long run, we expect the actual reuse of mapping to become a useful tool to rank mappings and marginalize mapping that are seldom reused.

6   Architecture, Data Model, and Scalability

The application is designed according to the traditional MVC design pattern, relying on J2EE JSF 2.0 framework[9] implementation of RichFaces 4.5.6 and MyFaces 2.23 for the Web interaction part. The mappings are also available through rest service, which are implemented relying on Jersey 2.6 framework[10]. The AJAX based user interface interaction grants quick response and easy interaction for the user. A view of the architecture of the application is presented in Fig. 6.

---

[6] http://dbpedia.org/ontology/Country.
[7] http://schema.org/AdministrativeArea.
[8] http://linkedgeodata.org/ontology/Country.
[9] http://docs.oracle.com/javaee/5/tutorial/doc/bnaph.html.
[10] https://jersey.java.net/.

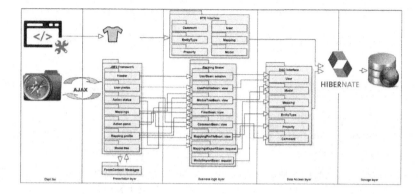

Fig. 6. A graphical view of the Synapsis architecture

Both the J2EE Backing Beans and the REST services access the mappings through standard Data Access Objects (DAO), which rely on the Hibernate ORM JPA 4.3.4 provider[11] to interact with a relational database (Oracle MySQL 5.5.27[12]) containing all data about mappings, users and supported models. A detailed view of the data model underlying the database is presented in Fig. 7.

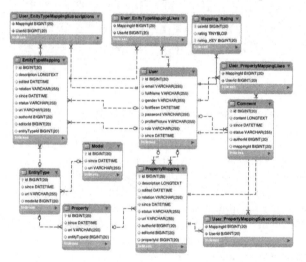

Fig. 7. Synapsis data model

6.1 Scalability Evaluation

We are proposing Synapsis as a web-application that should serve many users across several application scenarios. This is very likely to create a large number

---

[11] http://hibernate.org/orm/.
[12] https://www.mysql.com/.

Table 3. Scalability Test Numbers

Scalability test parameters	Number of items
Generated Models	181
Classes	4421
Properties	433671
Average number of classes per model	24
Average number of properties per model	2395
Tot. Mappings	109.000×10^6
Property Mappings	108×10^6
Class Mappings	1.1×10^6
Average number of mappings per property	249
Average number of mappings per class	250

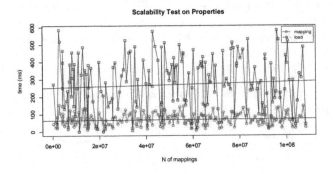

Fig. 8. Scalability test on property mappings

of mappings to be managed. Even though the amount of data will never count as big data, it is important to guarantee the application to be fully functional also when managing millions of mappings. Therefore, we set up a preliminary experiment with the goal of testing the scalability of the application managing a growing amount of vocabularies and mappings. In particular, our hypothesis is that the performance of the application measured in response time (wall clock time in ms) will not degrade while increasing the number of vocabularies and mappings. Hence, we decided to randomly generate a considerable amount of synthetic vocabularies and mappings across them to evaluate the degradation of performances measured in response time of some core operations (i.e. add mapping, load mapping) both reading and writing new information in the database. A description of the dataset of synthetic vocabularies mappings is presented in Table 3.

The creation of such dataset was performed iteratively, measuring at fixed intervals the response time in milliseconds of some basic operations (e.g. import a new model, create a mapping, like a mapping, add a comment, etc.) on a developer desktop machine. More precisely, we measured the interval of time

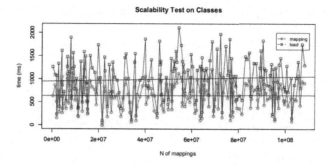

Fig. 9. Scalability test on class mappings

between the trigger of the execution of each operation and the full loading of the application response. All these operations have direct effect on the database, and therefore we are interested to measure the degradation of performance with an increasing amount of data to be managed. The test was implemented as a Java JDK 1.7.72[13] JUnit 4.11[14] program and executed simulating users' behavior by performing a sequence of operations while incrementing the number of models and mappings managed by the system. The machine on which the experiment was executed is a regular software developer desktop running on Linux Ubuntu 14.04LTS, processor Intel i7-3770 3.40 GHz with 16 GB of RAM and hard disk of 2 T 7.2 k rpm. The graph presented in Fig. 8 outlines the execution time for loading all the mappings for a target selected property and creating a single mapping for that property. The execution time measurements were taken computing the average of three repeated executions managing a range of mapping between 300k to 110M mappings. As shown in Fig. 8, the execution time for loading and creating property mappings grows very slowly with the increasing of the managed mappings, with an upper bound below 600ms for loading and near 150 ms for creating the mappings. A very similar pattern has been found for classes, as shown in Fig. 9. Although not outstanding, we consider these performances sufficient to support comfortably user interaction with Okkam Synapsis and therefore corroborate our experimental hypothesis. With the scalability experiments executed, we empirically tested the capability of the application in handling several millions of mapping without heavily affecting the user experience in terms of responsiveness. This test does not suffice to ensure that the application would scale up on a large number of concurrent users. However, relying on tools like Apache JMeter[15] we can easily find the performances degradation point, and react increasing redundancy of application server and database applying tradition scale-out best practices. In depth investigation of these aspects are negligible in this context. Future development and refinements will work on further improving these performances.

---

[13] https://www.java.com/en/.
[14] http://junit.org/.
[15] http://jmeter.apache.org/.

7   Kick-Off Mappings Dataset, and Licensing

Currently, Synapsis stores 22 mapping for equivalent class, and 205 mapping for subclasses of the entity type Person; 22 mappings for equivalent classes, and 2322 mappings for sub classes of the type Location; and finally we defined 20 mappings for equivalent classes and 2468 mappings for subclasses of the type Organization. These mappings were generated as contextual bridge rules to support semantic harmonization tasks in the knowledge-based solution described in [5]. In particular, the reader can find details about the process leading to the creation of such mappings from existing vocabularies towards the Identification Ontology[16] in Chap. 7 of [5]. We believe that this first core set of mappings can help to kick of a positive endeavor in the adoption of the Okkam Synapsis as a platform to create, share and manage mappings among vocabularies. Besides providing the first set of mappings, this is also the first example of application context in which Synapsis is involved as a user friendly tool to manage a consistent set of contextual real world mappings.

Any mapping created and shared through Okkam Synapsis is released under the very popular Common Creative Attribution 4.0 International license[17] (CC BY 4.0). The adoption of this copy-left license is to guarantee correct attribution and sharing without affecting the re-usability of the mappings (including commercial purposes). Therefore, contributors and consumers of mappings are granted what we believe is the ideal level of flexibility to accommodate requirements both of researchers and companies. Notice that all the mappings created through the application are subject to this license. Future evolution of the application may allow the selection of other licensing models to be compliant with loading of batches of mappings created else-where and under different licensing model including the share-alike.

8   Conclusion and Future Work

In this paper we have presented Synapsis, a platform which provides a gateway to collaboratively-defined ontology mappings. Looking for a pragmatic solution to the real world data heterogeneity, complexity and inconsistencies, we decided to enable users to define mappings as contextual bridge rules for their own applications, and enable peers to comment and discuss about them. We believe that definition of applications as context of use of mappings, and the relative of estimation *contextual sharedness* of mappings, will allow to filter commonly shared mappings, and at the same marginalize the long tail of over-specific ones. A beta version of the application is available at http://api.okkam.org/synapsis, and can be preliminarily tested and evaluated. We believe this tool can be suitable to support both Linked Data application developer and generic data curator to select the set of mappings of interest and have them available for the application through the defined rest services. We are also working on the development of a

---

[16] http://models.okkam.org/identification_ontology.owl ([5] Chap. 5).

[17] http://creativecommons.org/licenses/by/4.0/.

lightweight interface for editing single mappings without the need of uploading the target ontology/schema. In the near future, we will test the usefulness of contextual sharedness as a measure of relevance for the reuse of mappings, and introduce a system of rewards for the users based on traditional gamification approaches. Users will access badges, and upgrade their level from 'rookies' to 'gurus' based on the their level of contributions to the community and seniority. We believe these aspects are as important as smooth usability, and we will continue working on it to further improve and extend the functions available.

Acknowledgements. This work is partially supported by TAG CLOUD (Technologies lead to Adaptability and lifelong enGagement with culture throughout the CLOUD) FP7 EU Funded project, Grant agreement nr: 600924. This work has been partially funded by the Autonomous Province of Trento through SICRaS project (Legge 6/1999, DD n. 251).

References

1. Antoniou, G., van Harmelen, F.: A Semantic Web Primer. MIT Press, Cambridge (2004)
2. Berners-Lee, T., Hendler, J.A., Lassila, O.: The Semantic Web. Scientific American, May 2001. http://www.sciam.com/2001/0501issue/0501berners-lee.html
3. Berners-Lee, T.: Design Issues - Linked Data. Published online, May 2007. http://www.w3.org/DesignIssues/LinkedData.html
4. Bizer, C., Cyganiak, R., Heath, T.: How to publish linked data on the web. online tutorial, July 2007
5. Bortoli, S.: Knowledge Based Open Entity Matching. PhD thesis, International Doctoral School in ICT of the University of Trento (Italy) (2013)
6. Bortoli, S., Bouquet, P., Bazzanella, B.: An identification ontology for entity matching. In: Meersman, R., et al. (eds.) On the Move to Meaningful Internet Systems: OTM 2014 Workshops. LNCS, vol. 8842, pp. 587–596. Springer, Heidelberg (2011)
7. Bouquet, P., Stoermer, H., Niederee, C., Mana, A.: Entity name system: The backbone of an open and scalable web of data. In: Proceedings of the IEEE International Conference on Semantic Computing, ICSC 2008, pp. 554–561. IEEE Computer Society, August 2008
8. Bouquet, P., Giunchiglia, F., van Harmelen, F., Serafini, L., Stuckenschmidt, H.: C-OWL: contextualizing ontologies. In: Fensel, D., Sycara, K., Mylopoulos, J. (eds.) ISWC 2003. LNCS, vol. 2870, pp. 164–179. Springer, Heidelberg (2003)
9. Cohen, J.: A coefficient of agreement for nominal scales. Educ. Psychol. Measur. **20**(1), 37–46 (1960)
10. David, J., Euzenat, J., Scharffe, F., dos Santos, C.T.: The alignment appi 4.0. Semant. Web J. **2**(1), 3–10 (2011)
11. Ellis, J.B., Hassanzadeh, O., Srinivas, K., Ward, M.J.: Collective ontology alignment. In: OM, pp. 219–220 (2013)
12. Euzenat, J.: Uncertainty in crowdsourcing ontology matching. In: OM, pp. 221–222 (2013)
13. Fleiss, J.L.: Measuring nominal scale agreement among many raters. Psychol. Bull. **76**(5), 378–382 (1971)

14. Fu, B., Noy, N.F., Storey, M.-A.: Indented tree or graph? a usability study of ontology visualization techniques in the context of class mapping evaluation. In: Alani, H., et al. (eds.) ISWC 2013, Part I. LNCS, vol. 8218, pp. 117–134. Springer, Heidelberg (2013)

15. Gangemi, A., Presutti, V.: Ontology design patterns. In: Staab, S., Studer, R. (eds.) Handbook on Ontologies. International Handbooks on Information Systems, pp. 221–243. Springer, Heidelberg (2009)

16. Gangemi, A., Guarino, N., Masolo, C., Oltramari, A., Schneider, L.: Sweetening ontologies with DOLCE. In: Gómez-Pérez, A., Benjamins, V.R. (eds.) EKAW 2002. LNCS (LNAI), vol. 2473, pp. 166–181. Springer, Heidelberg (2002)

17. Alani, H., Correndo, G.: Collaborative support for community data sharing. In: Proceedings of The 2nd Workshop on Collective Intelligence in Semantic Web and Social Networks (2008)

18. W3C OWL Working Group. Owl2 web ontology language: Document overview, 27 October 2009. http://www.w3.org/TR/owl2-overview/

19. Guarino, N., Welty, C.: An overview of ontoclean. In: Staab, S., Studer, R. (eds.) The Handbook on Ontologies. International Handbooks on Information Systems, pp. 151–172. Springer, Heidelberg (2004)

20. Halpin, H., Hayes, P.J., McCusker, J.P., McGuinness, D.L., Thompson, H.S.: When owl:sameAs Isn't the Same: an analysis of identity in linked data. In: Patel-Schneider, P.F., Pan, Y., Hitzler, P., Mika, P., Zhang, L., Pan, J.Z., Horrocks, I., Glimm, B. (eds.) ISWC 2010, Part I. LNCS, vol. 6496, pp. 305–320. Springer, Heidelberg (2010)

21. Hitzler, P., Kroetzsch, M., Parsia, B., Patel-Schneider, P.F., Rudolph, S.: OWL 2 Web Ontology Language Primer (Second edition). W3C (2012)

22. Landis, R.J., Koch, G.G.: The measurement of observer agreement for categorical data. Biometrics **33**(1), 159–174 (1977)

23. Manola, F., Miller, E., McBride, B.: RDF 1.1 Primer. W3C, w3c working group note edition, June 2014

24. Markotschi, T., Völker, J.: Guess What?! human intelligence for mining linked data. In: Proceedings of the Workshop on Knowledge Injection into and Extraction from Linked Data (KIELD) at the International Conference on Knowledge Engineering and Knowledge Management (EKAW) (2010)

25. Noy, N.F., Griffith, N., Musen, M.A.: Collecting community-based mappings in an ontology repository. In: Sheth, A.P., Staab, S., Dean, M., Paolucci, M., Maynard, D., Finin, T., Thirunarayan, K. (eds.) ISWC 2008. LNCS, vol. 5318, pp. 371–386. Springer, Heidelberg (2008)

26. Patel-Schneider, P.F., Hayes, P., Horrocks, I.: Web Ontology Language (OWL) Abstract Syntax and Semantics. Technical report W3C, February 2003. http://www.w3.org/TR/owl-semantics/

27. Pavel, S., Euzenat, J.: Ontology matching: state of the art and future challenges. IEEE Trans. Knowl. Data Eng. **25**(1), 158–176 (2013)

28. Quine, W.: Ontological Relativity and Other Essays. Columbia University Press, New York (1969)

29. Raad, E., Evermann, J.: Is ontology alignment like analogy? - knowledge integration with lisa. In: Proceedings of Symposium on Applied Computing (SAC), Republic of Korea (2014)

30. Sarasua, C., Simperl, E., Noy, N.F.: CROWDMAP: crowdsourcing ontology alignment with microtasks. In: Cudré-Mauroux, P., et al. (eds.) The Semantic Web - ISWC 2012. LNCS, pp. 525–541. Springer, Heidelberg (2012)

31. Serafini, L., Zanobini, S., Sceffer, S., Bouquet, P.: Matching hierarchical classifications with attributes. In: Sure, Y., Domingue, J. (eds.) ESWC 2006. LNCS, vol. 4011, pp. 4–18. Springer, Heidelberg (2006)
32. Shvaiko, P., Euzenat, J.: Ten challenges for ontology matching. In: Meersman, R., Tari, Z. (eds.) On the Move to Meaningful Internet Systems: OTM 2008. LNCS, vol. 5332, pp. 1164–1182. Springer, Heidelberg (2008)
33. Thaler, S., Simperl, E., Siorpaes, K.: Spotthelink: playful alignment of ontologies. In: Proceedings of the ACM Symposium on Applied Computing, SAC 2011, pp. 1711–1712. ACM, New York, NY, USA (2011)
34. Tordai, A.: On Combining Alignment Techniques. PhD thesis, Vreije Universiteit Amsterdam, 03 December 2012
35. Tordai, A., van Ossenbruggen, J., Schreiber, G., Wielinga, B.: Let's agree to disagree: on the evaluation of vocabulary alignment. In: Proceedings of the Sixth International Conference on Knowledge Capture, K-CAP 2011, pp. 65–72. ACM, New York, NY, USA (2011)
36. Verborgh, R., De Wilde, M.: Using OpenRefine. PACKT Publishing, Birmingham (2013)
37. Zhdanova, A.V., Shvaiko, P.: Community-driven ontology matching. In: Sure, Y., Domingue, J. (eds.) The Semantic Web: Research and Applications. LNCS, vol. 4011, pp. 34–49. Springer, Heidelberg (2011)

Author Index

Printed in the United States
By Bookmasters